Understanding JCT Standard Building Contracts

Also available from Taylor & Francis

Construction Contracts Questions and Answers
D. Chappell   Paperback: 0–415–37597–5

Construction Contracts: law and management, fourth edition
J. Murdoch and W. Hughes   Hardback: 0–415–39368–X
                                              Paperback: 0–415–39369–8

Development and the Law: a guide for construction and property professionals
G. Bruce-Radcliffe   Hardback: 0–415–29021–X

Dictionary of Property and Construction Law
R. Hardy-Pickering   Hardback: 0–419–26100–1
                                              Paperback: 0–419–26110–9

The Presentation and Settlement of Contractors' Claims, second edition
M. Hackett and G. Trickey   Hardback: 0–419–20500–4

Information and ordering details

For price availability and ordering visit our website www.sponpress.com
Alternatively our books are available from all good bookshops

Understanding JCT
Standard Building Contracts

Eighth edition

David Chappell

Taylor & Francis
Taylor & Francis Group

LONDON AND NEW YORK

First published 1987 by International Thomson Publishing, reprinted 1988, 1989
Second edition published 1991 by E & FN Spon, reprinted 1992
Third edition published 1993, reprinted 1994
Fourth edition published 1995, reprinted 1996, 1997
Fifth edition published 1998, reprinted 1999 (twice)
Sixth edition published 2000, reprinted 2001, 2002 by Spon Press
Seventh edition published 2003 by Spon Press

This edition published 2007 by Taylor & Francis
2 Park Square, Milton Park, Abingdon, Oxon OX14 4RN

Simultaneously published in the USA and Canada
by Taylor & Francis
270 Madison Avenue, New York, NY 10016

Taylor & Francis is an imprint of the Taylor & Francis Group, an informa business

© 1987, 1991, 1993, 1995, 1998, 2000, 2003, 2007 David Chappell

Typeset in Sabon by Keystroke, 28 High Street, Tettenhall, Wolverhampton
Printed and bound in Great Britain by MPG Books Ltd, Bodmin

The publisher makes no representation, express or implied, with regard
to the accuracy of the information contained in this book and cannot
accept any legal responsibility or liability for any efforts or omissions
that may be made.

British Library Cataloguing in Publication Data
A catalogue record for this book is available from the British Library

Library of Congress Cataloging in Publication Data
Chappell, David.
  Understanding JCT standard building contracts / David Chappell.–8th ed.
    p. cm.
  Includes bibliographical references and index.
  1. Construction contracts–England. I. Title.
  KD1641.C488 2007
  343.41′07869–dc22   2006027544

ISBN 10: 0–415–41385–0 (pbk)   ISBN 13: 978–0–415–41385–5 (pbk)
ISBN 10: 0–203–96435–7 (ebk)   ISBN 13: 978–0–203–96435–4 (ebk)

Contents

Preface to the eighth edition

This little book continues to be popular among architects, quantity surveyors and contractors. I am gratified that it has been adopted as a standard text for students in schools of architecture and building as well as being read by those who are established in the industry. I will do my best to ensure that the text remains relatively simple and easy to read, and free from legalisms while remaining up to date and improved where possible. The original intention was to provide a straightforward guide to the three standard forms of contract in common use. In the last edition it was enlarged to deal with the most common form of design and build contract.

My guiding principle remains the kind of book I would have wanted when I was a newly qualified architect. What I wanted then and what was not available was a short book which told me all I needed to know about the then current forms of contract and which I could read without having to look up every other word in a legal dictionary. I wanted a book which was not too superficial, which gave me a few insights and which pointed the way to further reading.

All the JCT series of contracts and sub-contracts were completely rewritten and some new contracts were added during 2005. There are some notable omissions (nominated sub-contractors and suppliers), some additions (integral sectional completion and contractor's designed portion) and some substantial changes in structure and terminology, although the effect of the contracts remains much, but not entirely, the same as before.

This edition has been thoroughly updated to take account of these changes, and changes in legal decisions and relevant statutes. The JCT contracts dealt with in this book are:

- Standard Building Contract (SBC);
- Intermediate Building Contract (IC);
- Intermediate Building Contract with contractor's design (ICD);
- Minor Works Building Contract (MW);
- Minor Works Building Contract with contractor's design (MWD);
- Design and Build Contract (DB).

Although the book now covers two additional contracts (ICD and MWD), the two intermediate and two minor works contracts can be grouped together for most purposes.

As always, I am grateful to all those who have taken the trouble to express a view on the book, and all suggestions have been carefully considered and acted upon where appropriate.

Thanks as always to my wife, Margaret, for understanding that writing this kind of book must take precedence to mowing the lawn and household chores.

Note. Throughout this book, the contractor and any sub-contractor have been referred to as 'it' on the basis that they are corporate bodies.

David Chappell
David Chappell Consultancy Ltd
Wakefield, August 2006

Introduction

This book is written as a helpful general guide to six forms of contracts produced in 2006. These are the successors to the four popular forms of contract dealt with in the previous edition, namely the Standard Form of Building Contract with Quantities, 1998 (JCT 98), the Intermediate Form of Building Contract, 1998 (IFC 98), the Agreement for Minor Building Works, 1998 (MW 98) and the Standard Form of Building Contract 1998 with Contractor's Design (WCD 98). The book refers to contracts in England and Wales and substantially to Northern Ireland. It does not apply to Scotland, which has major legal and contractual differences.

It has been thought sensible to arrange the guide under a series of topics rather than undertake a clause-by-clause interpretation, because there are dangers in looking at the clauses in isolation and taking too literal an approach. So far as possible, the way in which all the contracts deal with a particular topic is examined together. However, it should be noted that:

- IC and ICD are dealt with together, as are MW and MWD. The differences between IC and ICD and between MW and MWD relate to the incorporation of the contractor's designed portion (CDP) in ICD and MWD. These differences are highlighted where they occur. Generally, for ease, unless the text dictates a different treatment, the four contracts are referred to in two groups as IC/ICD and MW/MWD.
- Provision for possession and completion in sections has been incorporated into SBC, IC, ICD and DB.
- SBC no longer has separate private and local authority versions, but it is still available for use with bills of quantities (SBC/Q), with approximate quantities (SBC/AQ) and without quantities (SBC/XQ).
- No architect is specified in DB. Any instructions or directions are given by the employer or, if appointed, the employer's agent. References in this book to the architect are never intended to mean an architect employed by the contractor. Such architects are treated as being part of the contractor's team – which indeed they are.
- References to case law have been inserted for the benefit of those who wish to read further and to show the way in which contract provisions

have been interpreted by the courts. It should be noted, however, that detailed expositions of cases have been excluded.

The book tries to state the law and the position under the contracts as at the end of August 2006.

Legal language has been avoided, and where the contractual position is obscure, a suggested course of action is laid down. In the interests of clarity, the provisions have been simplified; this book, therefore, supplements but does not take the place of the original forms. More than anything, it is intended to be practical, with emphasis on the contractor's interests. When in difficulty, the golden rule is to obtain expert advice.

1 Contractor's obligations

1.1 The forms

It seems appropriate to begin by looking briefly at the standard forms under consideration. All the JCT forms of contract were substantially amended in April 1998 to take account of the Housing Grants, Construction and Regeneration Act 1996 and the Latham Report. All the forms were reprinted at the end of 1998. JCT 80, IFC 84, MW 80 and CD 81 became JCT 98, IFC 98, MW 98 and WCD 98 respectively. In 2005 these were rewritten and became SBC, IC/ICD, MW/MWD and DB respectively.

SBC is a very comprehensive document which is suitable for use with any size of building works. Owing to its complexity, however, its use is likely to be reserved for projects which are substantial in value or complex in nature.

CD 81 (now DB) was introduced to produce a basis to allow contractors to carry out the design as well as the construction. In basic structure, and in some of the wording, it was based on JCT 80, and the resemblance between SBC and DB is still very strong. There lies the trap, because DB has many substantial differences to the traditional form of contract. Essentially, the scheme of the contract is that the employer, either personally or through an agent, produces a performance specification (the Employer's Requirements) which the contractor must satisfy. The contractor demonstrates how it intends to do this by producing the Contractor's Proposals. With this type of contract, the contractor carries most of the risk so far as cost, time and finished product are concerned. No independent architect is involved and, therefore, there are no certificates of any kind. There are merely statements and notices from the employer and applications for payment from the contractor. In addition to the normal clauses, this contract contains optional supplementary clauses.

IFC 84 (now IC/ICD) was introduced to fill the gap between JCT 80 (now SBC) and MW 80 (MW/MWD). A look inside the front cover suggests its use if the Works (all the work to be done) are of simple content, adequately specified or billed and without complicated specialist work. There is no suggested upper price limit, but £400,000 (at 2005 prices) and a maximum contract period of twelve months seem reasonable. Price and length of contract period are not, however, the most important factors.

MW and MWD are suitable for use on projects having a maximum value of £150,000 (at 2005 prices). They are not suitable for complex Works and no provision is made for bills of quantities or nominated sub-contractors. Very importantly, as far as contractors are concerned, there is only limited provision for reimbursement of loss and/or expense, although a claim can always be made using common law rights. This form is very popular, and not only for minor Works. It is known for it to be used in conjunction with bills of quantities, although quite unsuitable. The reason for its popularity is no doubt that it is short and simply expressed. Its simplicity is deceptive, however, and there are pitfalls for the unwary.

The contractor may think that the suitability or otherwise of a particular form for a particular project is academic in the sense that it can do very little about it. The choice is for the employer, advised by the architect. A thorough knowledge of the contents of the various forms, however, can influence the contractor's tender – if it has any sense.

Some employers use the standard forms, but with amendments to suit their own requirements and ideas. Such amendments, if substantial, may turn a standard form into the employer's 'written standard terms of business' under section 3 of the Unfair Contract Terms Act 1977 with the result referred to earlier. Amended forms of contract do have an unfortunate habit of back-firing on the party, making the amendments inconsistent or inoperative.[1] Any amendment to clauses 2.26–2.29 of SBC is likely to provide a bonus to the contractor unless great care is taken. More will be said about this later when dealing with extensions of time.

Whichever of these forms is used, the contractor undertakes to carry out the Works in accordance with the contract documents.

Contract documents

It is vitally important to know which are the contract documents, because they are the only documents which spell out what the employer and the contractor have agreed to do. Letters exchanged before the contract is entered into and the contractor's programme are not contract documents, i.e. they are not binding on the parties, unless expressly so stated. Architects may point to minutes of site meetings as evidence of what was agreed, but they cannot amend the contract documents.[2] In order to amend the terms of the contract it would be necessary for the employer (not the architect on the employer's behalf) and the contractor formally to agree the change, preferably in writing and preferably as a deed. SBC defines them in clause 1.1 as the contract drawings, contract bills, agreement, conditions and (if appropriate) the Employer's Requirements, Contractor's Proposals and the CDP (Contractor's Designed Portion) analysis. The contract drawings must be the drawings on which the contractor tendered. It is not unusual for the architect to have made revisions to the original drawings between tender and the signing of the contract. The contract drawings must be carefully scrutinised before signing and, if such revisions are present, the architect must be asked to restore them to their previous condition.

IC and ICD provide four options:

- contract drawings and specification priced by the contractor;
- contract drawings and work schedules priced by the contractor;
- contract drawings and bills of quantities priced by the contractor;
- contract drawings and specification and the sum the contractor requires for carrying out the Works;

together with the agreement and conditions (the printed form) and, if applicable, particulars of tender of any named person in the form of tender and agreement ICSub/NAM.

If the contractor is simply asked to state a sum required to carry out the Works – the last option – it must also supply a contract sum analysis or a schedule of rates on which the contract sum is based.

Strangely, neither the contract sum analysis nor the schedule of rates is a contract document. This may be important, because one of these two documents is essential to value architects' instructions requiring a variation (clause 5.3.1). It will normally be to the contractor's advantage if the third option is used (with priced bills), because it puts the onus on the employer to ensure that the quantities are correct. All the other options provide room for dispute if there are inconsistencies.

MW provides, in the second recital, and MWD in the third, for the contract documents to be any combination of contract drawings, specification and schedules together with the conditions (the printed form). MWD adds the Employer's Requirements. The third recital, the fourth in MWD, provides that the contractor must price either the specification or the schedules or provide a schedule of rates. In the latter case, the schedule of rates would not be a contract document.

DB defines the contract documents in clause 1.1 as the Employer's Requirements, the Contractor's Proposals and the Contract Sum Analysis together with the agreement and the conditions. The contents of the Employer's Requirements, the Contractor's Proposals and the Contract Sum Analysis are to be listed in the Contract Particulars. They are frequently composed of a mixture of specifications of various kinds and drawings. The Contract Sum Analysis is sometimes as detailed as bills of quantities.

It is usual to talk about 'signing' the contract, but in fact it can be executed in either of two ways: under hand (also known as a 'simple' contract) or as a deed (also known as a 'specialty' contract). As far as building contracts are concerned, there are two important differences. A deed does not need what is called 'consideration' to make it a valid contract, but a simple contract does need consideration. For example, a builder who offered to construct a house for someone would have to receive something in exchange for there to be a valid simple contract, but if the contract was entered into as a deed, it would be binding even if the builder agreed to build the house without any reward.

The second important difference concerns the Limitation Act 1980, which operates to limit the period during which either party may bring an action to enforce their rights under the contract. In the case of a simple contract, the period is six years from the date of the breach. For practical purposes, the starting date is usually taken as practical completion.[3] In the case of a deed, the period is twelve years. It is clear, therefore, that a contractor is more exposed if it enters into a building contract in the form of a deed.

It used to be the case that a deed had to be sealed in order to be properly executed. This was usually achieved by the impression of a device on wax or a wafer and fixed to the document. In fact, it was usually sufficient if it could be shown that both parties intended the document to be sealed.[4] The Law of Property (Miscellaneous Provisions) Act 1989 and the Companies Act 1989 have abolished the necessity for sealing for individuals and companies respectively. Indeed, sealing alone is not sufficient to create a deed. In Northern Ireland the Companies (No. 2) Order (Northern Ireland) 1990 and the Law Reform (Miscellaneous Provisions) (Northern Ireland) Order 2005 remove the requirement of sealing for companies and for individuals respectively. All that is necessary now for a document to be executed as a deed is that it must be made clear on its face that it is a deed and, in the case of a company, that it is signed by two directors or a director and company secretary, or, in the case of an individual generally, that it is signed by the individual in the presence of a witness, who must attest the signature. It is a matter on which proper advice should be sought before executing the contract.

Clause 1.8 of SBC, IC/ICD and DB allows the parties to agree that communications can be exchanged electronically, i.e. by e-mail, either by an insertion into the Contract Particulars or by written agreement. Security may be a factor to consider and the use of an electronic signature is sensible. It should be noted, however, that where there are specific provisions in the contract (e.g. instructions must be in writing), they must be observed. It is important that the other communications, which the parties agree may be sent electronically, are listed.

Discrepancies

It is quite usual for there to be some small, and sometimes large, discrepancies between the provisions in the printed form and in, say, the bills of quantities or specification. Priority of documents then becomes important. It is often thought that terms which are hand- or typewritten must take precedence over those which are printed because the written terms must represent the clear intentions of the parties. Indeed, that is the general law: type prevails over print.[5] However, clause 1.3 of SBC, IC/ICD and DB and clause 1.2 of MW/MWD clearly state that nothing contained in any of the contract documents will override or modify the terms in the printed form.

This kind of provision has been upheld in the courts.[6] In practice, it means that if a term in the contract bills or specification is in conflict with a term in the printed form, the printed term will prevail. For example, if the bills provide for an estate of houses to be completed on specific dates – in other words, phased completions – and the printed form contains just one date, it is the date in the printed form which will apply and the contractor will have fulfilled its obligations as to the time for completion if it completes all the houses on that one date. If the printed form stipulates that the period for payment is fourteen days, that stipulation cannot be overridden by a clause in the bills allowing twenty-one days. To be effective, the change must be made to the printed form itself.

All six contract forms are lump sum contracts. That is to say that, in general, the contractor takes the risk that the work may be more costly than he expects. Specific clauses, however, modify the effects.

In particular, when bills of quantities are used, SBC clause 2.13.1 and IC/ICD clause 2.12.1 expressly provide that the bills are to be prepared in accordance with the Standard Method of Measurement (SMM) unless specifically stated otherwise in respect of particular items. Errors in the bills are to be corrected and treated as variations. The effect of this is that a contractor is entitled to price everything as though measured in accordance with SMM unless there is a note on that item. A general note to the effect that not everything is measured in accordance with SMM would not be effective. Contractors can recover substantial amounts of money simply by paying attention to this point.

Under DB, there is less scope for the contractor to claim additional costs, because it is generally taken to have allowed in its price for satisfying the Employer's Requirements.

If the contractor makes an error in pricing which is not detected before acceptance of the tender, it is stuck with it. This may seem harsh, and many *ex gratia* claims are made on this basis, but if two parties contract together they are usually taken to know what they are doing. It may be possible for the contractor to obtain relief if it can show that the employer detected a substantial mistake in pricing and, knowing that the contractor would not wish to contract on those terms, purported to accept the tender.[7] In Canada, an architect has been held liable to a contractor for failure to include important information in the invitation to tender. As a result, the contractor was unable to use its preferred system and lost money.[8]

It is now established that, under SBC, IC/ICD and MW/MWD, the contractor has no duty to search for discrepancies and inconsistencies between the contract documents. The architect is responsible for providing the contractor with correct information. If the contractor does not discover a discrepancy until too late, the employer must pay for any additional costs resulting.[9] In the nature of things, inconsistencies will be present. All the three forms make provision for the correction of such inconsistencies: SBC in clause 2.15, IC/ICD in clause 2.13, MW in clause 2.4 and MWD in clause 2.5.

The architect may well consider that a contractor which is carrying out its obligations properly will have to examine the documents carefully and so detect discrepancies. This favourite argument does not change the legal position noted above. Of course, upon finding a discrepancy, the contractor should always ask the architect, in writing, for an instruction.

Most of the contractor's problems in this respect arise because it is anxious to proceed 'regularly and diligently', and when confronted by two drawings, or a drawing and a bill item, which do not correspond, it attempts to solve the problem itself. By so doing, the contractor loses its right to payment for any variation and probably takes responsibility for the design of that particular part of the work. If, however, the contractor does not proceed on that particular part of the work, asks the architect for instructions and notifies delay and disruption, it will be entitled to the whole of its loss, in terms of time and money, as well as payment for the variation instruction when it eventually arrives.

A closely linked situation, although not strictly an inconsistency, is where the contractor is not provided with a particular detail it requires. The contractor may think it knows what the architect intends to be done, but if the contractor carries out without precise instructions, it will probably become liable for any inadequacy in the detail.[10] The contractor should deal with it in precisely the same way as inconsistencies.

Although payment for architect's instructions requiring a variation to correct a discrepancy will be fairly automatic under the terms of SBC, the same may not be true of IC or ICD, where bills are not used, and MW/MWD.

For example, under the provisions of clause 4.1 of IC/ICD, the contractor will be deemed to have priced for work in the specification even if not quantified. The rules in this clause are quite complicated and will repay careful study before tendering. Broadly, the rule is that if there are no quantities for a particular item, the contract documents must be read together. If there is a conflict between documents, the drawings prevail. Where quantities are shown, they prevail.

One of the dangers of the deceptively simple MW and MWD is that the contractor's obligation is to carry out the Works in accordance with the contract documents on which it has submitted its price. The contract documents will probably be drawings and specifications, and therefore the contractor will be deemed to have included for the whole of the work shown on the drawings and specification. If there is an inconsistency, it appears that the contractor must be deemed to have priced for either option. In consequence, it appears that cases where the contractor will be entitled to additional payment as a result of inconsistencies will be rare.

In view of the different philosophy underlying DB, discrepancies are dealt with in an entirely different way. Two situations are envisaged: a discrepancy within the Employer's Requirements and a discrepancy within the Contractor's Proposals. In each case, employer and contractor share the duty of informing the other if either discovers a discrepancy. A discrepancy in the Contractor's Proposals is covered by clause 2.14.1. The contractor must suggest an amendment and the employer may choose between the discrepant items and the suggestion at no additional cost. Under clause 2.14.2, a discrepancy in the Employer's Requirements is dealt with in whatever manner is stated in the Contractor's Proposals or, if not so stated, as suggested by the contractor, which the employer can either accept or reject in favour of his or her own solution. Either way, it is to be treated as a change (which is the DB term for a variation).

A topic which frequently causes difficulty is when there is a discrepancy between the Employer's Requirements and the Contractor's Proposals. The contract makes no provision to deal with this situation. Indeed, one of the footnotes emphasises the importance of removing all discrepancies between the two documents. We all (except, it seems, the JCT) know that life is not like that and discrepancies will occur. The straightforward way of resolving such matters is on the basis of priority of documents.

Although it is not expressly stated, the contract wording clearly points to the Employer's Requirements taking precedence over the Contractor's Proposals. The third recital of the terms and conditions provides that the employer has examined the Contractor's Proposals and is satisfied that they appear to meet the Employer's Requirements. It is clearly not intended that, under the contract, the employer or the employer's advisers should check the Contractor's Proposals exhaustively to *ensure* that they meet the Employer's Requirements. Had such a thing been intended, it would have been easy for the draftsman to have used clear words to that effect. However, the wording strongly points to the intention that the Contractor's Proposals will be drafted to meet the Employer's Requirements.

Because the contract does not expressly deal with discrepancies between Employer's Requirements and Contractor's Proposals, it is likely that a court would imply a term that is necessary for the business efficacy of the contract – giving precedence to the Employer's Requirements. This is because:

- The contract philosophy is that the employer sets out requirements. The intention is that the Employer's Requirements and the Contractor's Proposals should dovetail together and that, where they do not do so, that is a qualification or divergence. It would be perverse to permit the Proposals to take precedence. The employer is entitled to assume that the contractor is complying with the Employer's Requirements.
- The Contractor's Proposals should be an indication of how the contractor is to comply with the Employer's Requirements – not an indication of how it wishes to construct the project or allocate risk. The wording of the first and second recitals reflects this.

- Clause 2.2 provides that the Employer's Requirements prevail over the Contractor's Proposals where workmanship or materials are concerned.
- Under the terms of the contract, the employer cannot issue a change instructing the contractor to vary the Contractor's Proposals. Clause 5.1.1 provides that a change means a change in the Employer's Requirements. Neither can the employer instruct the expenditure of a provisional sum in the Contractor's Proposals (see clause 3.11). In the absence of a right to instruct a change to, or the expenditure of a provisional sum in, the Contractor's Proposals it would be perverse for the Contractor's Proposals to prevail over the Employer's Requirements, because that would dis-entitle the employer from issuing changes in respect of the discrepant parts of those Contractor's Proposals.

These points generally apply to the Employer's Requirements and Contractor's Proposals in the CDPs of SBC and ICD. The position under MWD is different in detail, because although there is provision for Employer's Requirements there is none for Contractor's Proposals. The situation here is even clearer. Article 1 requires the contractor to comply with the contract documents, one of which is expressly stated to be the Employer's Requirements. Therefore, under MWD, there can be little doubt that the contractor must comply with the Employer's Requirements.

1.2 Implied and express terms

The general law will imply three important terms into all building contracts:

- The contractor will carry out the work in a good and workmanlike manner.
- The contractor will supply and use good materials.
- The contractor will undertake that the completed building is reasonably fit for its intended purpose where that purpose is known and there is no other designer involved.

These terms can be modified or supplanted only if there are express terms (i.e. written in) in the contract dealing with the same topics or, in the case of the third term, if someone has been employed in a design capacity.

Implied terms come as a shock to some contractors, who think that the whole of their obligations are covered by the terms which can be read in the printed document. There are, in fact, a good many other terms which the law will imply.

Some of these terms are implied by statute, for example the Supply of Goods and Services Act 1982 and the Defective Premises Act 1972. In addition, the Unfair Contract Terms Act 1977 will often operate to void the effect of some clauses in a contract, particularly those clauses seeking to exclude or reduce liability. Where consumers are concerned, the Unfair Terms in Consumer Contracts Regulations 1999 strictly protect the rights of consumers, and clauses which have not been negotiated may be regarded as unfair.

The courts are usually reluctant to imply terms into a contract made between two parties, because, provided that there is no mistake, misrepresentation, illegality, etc. in a contract, it is generally considered that the parties should be left with the bargain they have made, no matter how uncomfortable it turns out to be. When the courts do imply a term into a contract, it must not be inconsistent with an express term and it must have as its basis the presumed intentions of the parties; in other words, what they would have written into the contract had they given thought to the matter before the contract was made. The courts, however, will not imply terms into a contract simply because it seems to be a good idea. Terms will be implied only in accordance with certain principles which the courts themselves have built up over a number of cases. In general, the courts will imply a term if:

- it is the kind of term which, if asked, both parties would immediately agree was part of their bargain;
- the contract is apparently complete, but lacks just one term to make it workable;
- the parties have not fully stated the terms and the court is seeking to define the contract;
- in all the circumstances, it is reasonable to do so (rarely).

It can readily be seen that these categories are not completely separate, but rather instances of differing emphasis.[11]

In SBC, IC/ICD, MW/MWD and DB the contractor's primary obligations are stated in clause 2.1 in each case. It is no accident that these obligations are at the very beginning of the conditions. In each contract the obligations are expressed in very similar terms: the contractor is to carry out and complete the Works in accordance with the contract documents. This is a basic and absolute obligation from which the contractor can expect relief only in very restricted circumstances; for example, if the employer prevents completion,[12] or if the contractor lawfully terminates its own employment in accordance with the appropriate clause (SBC, IC/ICD and DB clauses 8.9, 8.10 and 8.11, MW/MWD clauses 6.8, 6.9 and 6.10). It is considered that generally this clause is not sufficient to impose any design liability on the contractor.[13] But it should be noted that under DB the contractor is made responsible for completing the design of the Works and under ICD and MWD the contractor has design responsibility for a portion of the Works.

Contractor's duties

The contractor's duties are scattered throughout each contract. There are over ninety separate instances in SBC, more than seventy in IC/ICD and twenty-three/twenty-nine in MW/MWD respectively. The contractor has over seventy duties under DB. These, it must be remembered, are express duties and take no account of the duties which will be implied. It is, therefore, of the utmost importance that the contractor reads the contract carefully. It should be considered a working tool, not something to be thrown into a drawer. It is well worth while highlighting the duties with a coloured pen.

Very often, the contractor's entitlement to money or an extension of time depends upon the correct performance of some obligation as a precondition. A good example is to be found in SBC clause 4.23, where a precondition to the ascertainment of loss and/or expense is that the contractor must make application as soon as it has become, or should reasonably have become, apparent that the regular progress of the Works has been or is likely to be affected. Failure to carry out the duty precisely could seriously prejudice the contractor's chances of obtaining reimbursement. Thus, a contractor that waits until several weeks after the event might well be considered to be in breach of its duty.

An obligation which seems to cause more difficulty is contained in SBC and IC/ICD clause 2.4 and DB clause 2.3. This is the obligation to proceed with the Works regularly and diligently. MW/MWD does not expressly refer to carrying out the Works regularly and diligently but, strangely, failure to do so is made a ground for termination. The question is, what does the obligation mean in practice?

The words will be interpreted in the light of the facts, but it seems clear that simply going slow will not necessarily be taken to mean that the contractor is not proceeding diligently. It has been suggested that an obligation to work with due diligence is an obligation on a contractor to work so as to meet the key dates and the completion date in the contract.[14] It has been established that the duty is to proceed regularly and diligently (the words being read together), efficiently and industriously, using the resources necessary to complete the contract on time. It is not sufficient for the contractor to have men on site at regular intervals; the work must be being progressed.[15] This is important in view of the fact that failure to proceed regularly and/or diligently is a ground for termination by the employer in all six contracts. Failure on the part of the contractor to keep to its programme would hardly qualify under this head unless such failure was gross. But an architect who did not issue a default notice in the face of the contractor's clear failure to proceed regularly and diligently would be in breach of duty to the employer. The court of appeal has explained the meaning of regularly and diligently as follows:

> What particularly is supplied by the word 'regularly' is not least a requirement to attend for work on a regular daily basis with sufficient in the way of men, materials and plant to have the physical capacity to progress the Works substantially in accordance with the contractual obligations.

What in particular the word 'diligently' contributes to the concept is the need to apply that physical capacity industriously and efficiently towards the same end.

Taken together the obligation upon the contractor is essentially to proceed continuously, industriously and efficiently with appropriate physical resources so as to progress the Works steadily towards completion substantially in accordance with the contractual requirements as to time, sequence and quality of work.[16]

SBC clause 2.9.1.2 requires the contractor to provide two copies of its master programme and to update it within fourteen days of any decision by the architect in relation to extensions of time. IC/ICD, MW/MWD and DB have no such provision, but there is no reason why such a requirement should not be incorporated into the specification or preliminaries. It is useful if the programme is specified as being in network form, because it enables the estimation of extensions of time much more easily than with a simple bar chart. The courts have approved the use of computerised techniques to input the delays into a precedence diagram to determine the likely extension of time.[17] All architects and contractors should make use of this facility.

Contractors often show on their programmes that they intend to complete several weeks before the date for completion in the contract. It is the contractor's privilege to complete earlier than the date for completion if it wishes, but the employer, through the architect, is not bound to assist the contractor by providing information earlier than is necessary to enable the contractor to complete the Works by the contractual date.[18] It may not be to the contractor's advantage to produce a programme showing that it will finish four weeks early. The architect may argue that the contractor is not entitled to any extension of time until delays affect the end date by more than four weeks. Of course, in such an instance the contractor would be entitled to recover whatever direct loss and/or expense it could prove if matters giving rise to loss and/or expense were the cause of the delay.

Architect's satisfaction

The contractor's basic obligation is qualified in each of the forms under consideration, except DB, by the proviso that where and to the extent that the approval of quality of materials or standards of workmanship is a matter for the opinion of the architect, it must be to the architect's reasonable satisfaction. This does not impose a dual responsibility upon the contractor unless the contract documents expressly state that is to be the situation.[19]

In general, the contractor must follow the requirements laid down in the contract documents, such as to provide timber to a certain standard. Provided the contractor does just that, it has fulfilled its obligations. It matters not that the architect is not satisfied, provided the materials or workmanship are in accordance with what is laid down in the contract. The architect may not be satisfied and simply disappointed with the results of the specification, but if the architect wants an improved result in such a case, the employer has to pay for it.

If, however, materials or workmanship are stated in the contract documents to be to the architect's satisfaction, or some such phrase, the contractor's obligations will be to reasonably satisfy the architect. The architect cannot specify iron and expect gold. In effect, the architect's approval will override any specification requirement. It is, therefore, important to obtain such approval in writing.

Even if the contractor does not manage to ensure that the architect expresses approval in writing on every occasion that the architect expresses it orally, the issue of the final certificate, in the case of SBC, IC and ICD, will be conclusive evidence that such approval has been given (SBC, IC and ICD clause 1.10.1.1). The position is similar under DB regarding the employer's approval, except that there is no final certificate; merely a final account and final statement which may become conclusive about the same kind of things as under SBC. The final certificate under MW/MWD is not conclusive about anything.

This is of crucial importance, especially to those contractors who may be worried at tender stage about the frequency of architect's approvals in the specification. Of course, it does not remove the obligation to satisfy the architect, but once the architect is satisfied and the final certificate has been issued, it becomes the architect's worry.

In 1994 it was established that quality and standards of materials and workmanship are always matters for the opinion of the architect.[20] The effect was that the final certificate under the then JCT 80 or IFC 84 was conclusive that the architect was satisfied with these matters whether or not they had been mentioned in the specification or bills of quantities. Clause 1.10.1.1 of SBC, IC and ICD and clause 1.9.1.1 of DB in respect of the final account have been amended by the JCT with the object of restoring the situation to the pre-1994 position. It should be noted that, in each case, the final certificate is expressly stated not to be conclusive that any materials, goods or workmanship comply with the contract.

If the architect does not issue the final certificate at the correct date, the employer will not be able to take advantage of the failure in order to take action against the contractor for defects if the action would have been prevented had the final certificate been correctly issued.[21]

SBC clause 3.20 stipulates that where materials, goods or workmanship are to be to the architect's satisfaction, any dissatisfaction must be expressed within a reasonable time of the work being carried out. In the light of the remarks above, it seems that clause 3.20 extends to all materials, goods and workmanship whether or not expressly noted in the specification or bills of quantities. This is an added safeguard for the contractor against the architect waiting until the project is nearly complete before making a complaint known. It puts the onus on the architect to decide whether or not he is satisfied. The precise meaning of 'a reasonable time' will always be open to dispute, but in this context it is thought that the architect must express any dissatisfaction before the contractor carries out the next operation. For example, if the contractor lays a screed, the architect must express any dissatisfaction before the floor finish is applied and possibly before all wet trades have finished. Employers will no doubt delete this provision.

Duty to warn

Although the position is not crystal-clear, it seems that the contractor has no duty to warn the architect of defects in the architect's design, although, in certain circumstances, the contractor may have a duty to warn the employer if it can be shown that the employer is placing special reliance upon the contractor.[22] The contractor certainly has a duty to the employer in those cases where the architect has produced the original drawings but is taking no further part in the project.[23] The prudent contractor will always notify the architect if it considers that there is a design defect, irrespective of whether it has a duty to do so.[24] It is wise to put the notice in writing. The architect who ignores such a warning would be misguided, to say the least, but then no blame can attach to the contractor. There are occasions when there is a danger of death or injury where it seems that no amount of warning on the part of the contractor will suffice and nothing short of refusal to proceed will discharge its responsibility.[25]

1.3 Design

In general, the contractor will have no obligation to, or liability for, design under a so-called traditional contract unless it has taken it upon itself (not uncommon), or unless the contract documents clearly set out such an obligation and liability. Many architects mistakenly thought that the former IFC 98 gave the contractor some design responsibility. That was never correct. The general position is now modified by SBC, ICD and MWD, all of which provide for a Contractor's Designed Portion incorporated as part of the contract. There is no longer any necessity to use additional supplements. Architects often specify items which unavoidably involve design, and the contractor secures the items by means of sub-contractors or suppliers. Problems arise when design defects appear. The architect should now use the CDP when the supply of such things as roof trusses, precast floor beams or any other element having a design content and which the architect does not wish to design is required. It is convenient to consider the CDP in each contract as a little design and build section. Under SBC and ICD the employer prepares a performance specification in the form of Employer's Requirements to which the contractor responds with its Contractor's Proposals. Both documents become part of the contract documents together with a breakdown of the part of the contract sum related to the CDP work known as the CDP Analysis. Discrepancies are dealt with broadly in the same way as under DB.

DB, ICD and MWD clause 2.1 and SBC clause 2.2 require the contractor to complete the design of the CDP (the whole of the Works in the case of DB, of course). It is the contractor's responsibility for completion of the design of the *whole* of the Works which distinguishes DB. Although the responsibility appears to be confined to completing the design, it has been held that, in carrying out design and build, the contractor has the duty to review whatever design is presented to it and to ensure that it works.[26] This is, of course, quite an onerous responsibility. The latest contracts seek to overcome this judgment by the insertion of a clause which provides that the contractor is not to be liable for anything in the Employer's Requirements nor for any design in the Requirements (SBC clause 2.13, ICD clause 2.34, MWD clause 2.1.2 and DB clause 2.11). This has the effect of confining the contractor's responsibility to completing the design. Obviously, a contractor who can see that an existing design is seriously defective probably has a duty to warn (see section 1.2 above).

Under clauses 2.19.1, 2.34.1, 2.1.1 and 2.17.1 of SBC, ICD, MWD and DB respectively, the contractor's design liability is confined to reasonable skill and care, just like an architect. This is an important clause, because without it the contractor's liability would be the higher standard of fitness for purpose.[27] The architect has the power to issue directions for the integration of CDP work with the rest of the design.

Only SBC (clause 2.41), ICD (clause 2.33) and DB (clause 2.38) include a clause to deal with copyright. Although the contractor holds the copyright in each case, the employer is given an irrevocable licence to use the contractor's design in connection with the Works. It should be noted that there is an important proviso. The licence is subject to all money due and payable to the contractor having been paid. On this basis the employer does not have a licence until the building is finished and the final certificate issued and paid. In order to make this clause workable, the employer may perhaps be considered to have a licence throughout the progress of the work, so long as payment is made of money due, and that the licence becomes irrevocable only on payment of the final amount. This clause needs more thought and probably some redrafting. Although there is no express clause dealing with copyright under MWD, the general law would apply, in particular the Copyright, Designs and Patents Act 1988. Under the general law, copyright would remain with the contractor as creator of the design and the employer would acquire a licence to reproduce the design in the form of a building provided that a significant amount had been paid to the contractor in respect of the design.[28] That would be likely to fall somewhat short of payment of the whole of the final account figure.

Only DB and SBC include provision for the submission of design drawings by the contractor. They are contained in schedule 1 in each case. The procedure is straightforward, but repays careful reading. The architect (or the employer under DB) must return one copy of each drawing marked either A, B or C. The contractor may proceed with drawings marked A. Drawings marked B may be used provided the contractor strictly adheres to the comments. Drawings marked C must be resubmitted. Submission of the drawings cannot be taken as an opportunity to amend the design. Drawings may be marked B or C only if they are not in accordance with the contract. For example, the architect has no power to require the contractor to change something on a drawing just because the architect would have done it a different way. In order to mark drawings B or C, the architect (or the employer under DB) must be able to say that the drawings did not comply with the Employer's Requirements or, if not shown on the Employer's Requirements, with the Contractor's Proposals. Probably comment can also be made if it can be shown that the contractor's construction detail would not work.

1.4 Materials and workmanship

SBC, IC, ICD, MW, MWD and DB clause 2.1 impose substantial obligations on the contractor to carry out and complete the Works in accordance with the contract documents. Moreover, items for the architect's approval, so far as quality and standards are concerned, are to be to the architect's reasonable satisfaction, but not of course under DB. These obligations form a basis against which the contract clauses dealing with materials and work-manship should be studied.

MW makes no further reference to materials, goods and workmanship except that materials and goods reasonably and properly brought on to site and adequately stored and protected are to be included in progress payments (clause 4.3). MWD, into which the CDP provisions have been grafted somewhat awkwardly, includes a set of provisions in clause 2.2.1, most of which are contained in MW clause 2.1.

Procurable

A very useful protection is given to the contractor by SBC clause 2.3.1 and DB clause 2.2.1. The contractor's duty is stated to be to provide materials and goods of the standards described 'so far as procurable' – that is, obtainable. If the contractor cannot obtain goods, etc. of the standards described, its obligation seems to be at an end.

The clause provides no escape for a contractor who is finding it more difficult or more expensive to obtain the required standard of goods. Unexpected rises in prices are the contractor's risk. If the materials, etc. are unobtainable because the contractor was late in placing the order, it is probable that the contractor has a duty to supply materials, etc. at least equal in standards to what is specified and at no extra cost to the employer.

If the materials become unobtainable after the contractor's tender is accepted, it then falls to the architect to issue an instruction to vary the materials or, under DB, for the employer to consent to the substitution of another material. If the new materials prove more costly, the contractor is entitled to be paid the difference. Neither IC, ICD, MW nor MWD provides this relief for the contractor, who will be responsible under these contracts for providing alternatives at the original cost. If the amount of unobtainable material is substantial, it may amount to frustration of the contract.[29]

Workmanship has been separated from materials and goods under SBC. So far as workmanship is concerned, it must be of the standard described in the bills of quantities (or specification if the Without Quantities version is being used). Both workmanship and materials under SBC must be to the reasonable satisfaction of the architect if the architect so requires in accordance with clause 2.3.3. Where the CDP is being used, the workmanship must be as described in the Employer's Requirements. Only if not described in the Employer's Requirements is the workmanship to be as described in the Contractor's Proposals.

Under DB, materials and workmanship must be in accordance with the Employer's Requirements. It is only if they are not specifically described in the Employer's Requirements that the contractor is entitled to look at its own Proposals or at any document it has subsequently provided under clause 2.8. This is extremely significant and indicates that the prime document is the Employer's Requirements.

SBC, IC, ICD, MW, MWD and DB clause 2.1 provide that all work must be carried out in a workmanlike manner. SBC clause 3.19 and DB clause 3.14 provide that if the contractor fails to comply with this provision, the architect (or the employer under DB) may issue whatever instructions may be reasonably necessary, including the instructing of a variation, in consequence. First the contractor must be consulted. Provided, and to the extent that, they are necessary instructions, the contractor is not entitled to any payment for the variation, or to any extension of time. This clause is something of a 'catch-all' attempt to cover those circumstances where it cannot be said that the work is not in accordance with the contract but it has not been carried out in a workmanlike manner so that it is dangerous or it threatens the proper carrying out of other parts of the Works.

Opening up and testing

The contractor may be required to open up work already covered up or to carry out testing of materials, even if they are already built into the Works (SBC clause 3.17, IC/ICD clause 3.14 and DB clause 3.12). MW and MWD make no specific reference to opening up and testing, but the power to order such work probably exists under clause 3.4. The contractor is entitled to be paid for opening up and making good again and for any tests required unless:

- the cost of such opening up and testing is already included in the contract;
- the work or materials are found to be not in accordance with the contract.

If it is suggested that the cost is already included in the contract, the contractor should make sure that it is indeed included. It seems unlikely that any very general note in the specification would cover the situation, and the amount of opening up and testing would have to be specified in reasonable detail. Otherwise, the contractor could be required to undertake limitless amounts of investigative work of this nature with no recompense. If uncovered work is found to be defective, the contractor has to bear the cost of uncovering, correcting the defects and making good again. That appears to be reasonable.

Contractors often feel aggrieved when asked to open work up. The grumble seems to be threefold: it throws doubt on their competence, the architect or clerk of works could have inspected before covering up and the contractor feels that, after opening up, some excuse will be found to justify the exercise and leave the contractor with the expense. Thus, if opening up is instructed in order to inspect part of the foundations, just before practical completion of the whole building is due, the contractor will be understandably irritated.

Contractors just have to live with the fact that they are not always trusted. Opening up is often ordered because some aspect of the work provokes the suspicion that all is not well. Take, for example, the case of a specified 60 mm concrete screed laid on a solid concrete ground floor with a separating membrane between. Severe cracking may be noticed and it may be suspected that the screed is not thick enough, and instructions may be given for opening up of part of it to make sure. Testing of pieces of the removed screed may also be ordered to check that it is of the correct mix, etc. If it is found that the screed is at least 60 mm thick and of the correct mix, etc., and the separating membrane is in position and undamaged, the contractor would have a good case for reimbursement.

Undoubtedly the screed is defective in the generally accepted sense because it is badly cracked, but it is demonstrably in accordance with the contract, which is all the opening up clause requires for the contractor to obtain payment. The problem may be that, in the circumstances, the architect should have specified a reinforced screed. On the other hand, if everything is found to be in accordance with the contract except that the screed is only 55 mm thick, it will avail the contractor nothing to protest that the 5 mm difference cannot possibly have caused the cracking. The contractor may argue that, to avoid cracking, some light reinforcement would be necessary, that cracking, in itself, is not serious and so on, but it will still be expected to remove the screed and re-lay it to the required thickness at no extra cost. If, however, in the process of re-laying, the architect instructed the inclusion of additional light reinforcement in the new screed, the contractor would have an arguable case that it should be paid for more than the additional reinforcement. However, under DB, or if it is the subject of a CDP in SBC, ICD or MWD, a design or specification error will usually also be a matter for the contractor.

It should be noted that the architect has no duty to the contractor to find defects.[30] The architect's duty is to the employer. The contractor can have no recourse against the architect who does not detect a defect until late in the contract, because it is the contractor's obligation to build in accordance with the contract documents.

Failure of work

An important provision in IC/ICD (clause 3.15) relates to failure of work. If any materials or work are found to be at variance with contract requirements, the contractor must, without prompting, tell the architect how it intends to ensure that there are no similar failures elsewhere on the job.

The proposed measures must be entirely at the contractor's own cost. For example, if it is discovered that the dpc has been omitted over some windows, the contractor might propose opening up, say, 10 per cent of all window heads for inspection by the architect. If the contractor's proposal is accepted, the contractor has to stand the cost even though all inspected window heads have dpcs in accordance with the contract. The contractor would, however, be entitled to an appropriate extension of time. The contractor must submit its proposal within seven days of the discovery of the initial defect.

If the architect is not satisfied with the contractor's proposal, if the proposal is not submitted within seven days or if there are some pressing safety or statutory reasons why a wait of even seven days is too long, the architect may issue instructions for opening up, as the architect sees fit, at the contractor's expense. In such a situation the contractor has ten days in which to decide whether to carry out the instruction or to write to the architect with its objections. Objections will usually relate to the amount of opening up required, but there could be other grounds. If the architect does not withdraw or modify the instruction within a further seven days, the matter must be referred to one of the dispute resolution procedures.

A problem for the contractor lies in the fact that, in this clause, no distinction is made between major and minor failures of work. The only qualification is that the proposals are restricted to establishing that there are no similar failures. On every contract there will be a multitude of very minor instances, readily corrected, where work is not in accordance with the contract and one or two cases which may be major. The contractor would be well advised to settle with the architect, at an early stage, in what circumstances the architect expects clause 3.15 to be operated. This is best done at the first site meeting and recorded in the minutes, or, less easily, by an exchange of letters.

SBC clause 3.18.1, IC/ICD clause 3.16.1 and DB clause 3.13.1 empower the architect (or the employer under DB) to order the removal from site of work or materials which are not in accordance with the contract. MW and MWD have no similar provisions, but it is thought that the architect must have this power under clause 3.4. Note that an instruction to correct defective work is not valid (except possibly under MW/MWD terms). The instruction must require removal of the work from the site.[31] In practice, of course, it usually amounts to the same thing.

Under SBC and DB, if materials, goods or workmanship are not in accordance with the contract, the architect (or the employer under DB) may do any or all of the following:

- Instruct that the materials, etc. be removed from site.
- Allow the work to remain and make an appropriate deduction from the contract sum. The employer must agree and the contractor must be consulted.
- Issue such instructions requiring a variation as are reasonably necessary as a consequence of previous action under this clause. There is to be no additional cost, extension of time or loss and/or expense. Once again, the contractor must be consulted.
- Issue instructions under clause 3.18.4 or 3.13.3 under SBC or DB respectively requiring the contractor to open up or test the work to establish to the reasonable satisfaction of the architect (employer under DB) the likelihood of any other similar instances of failure to comply with the contract. If the instruction is reasonable, the contractor is not entitled to any addition to the contract sum, but it is entitled to an extension of time unless work is found not to comply with the contract.

In using the last power, 'due regard' must be had to a code of practice. The code is part of the printed contract form and it is a list of relevant factors required to be considered in order that the extent of instructions to open up is reasonable. There are fifteen items, ranging from the need to show the employer that a particular defect does not occur throughout the Works to proposals the contractor may make and 'any other relevant matters' – which just about covers everything. There is no requirement that the instruction to the contractor must be justified. Reference to the code will probably be most useful if the contractor decides to seek adjudication or arbitration on whether its objection to the instruction is justified.

There is a danger for the employer if the instruction states how defective work is to be remedied. It is sometimes thought (quite wrongly) that because the contractor is in breach of contract in providing materials or workmanship not in accordance with the contract, removal from the site and replacement with more expensive work or materials can be ordered. If such instructions are given, the contractor is entitled to be paid as though a variation were being ordered.[32]

A question which sometimes arises on traditional contracts such as SBC, IC, ICD, MW and MWD is the extent to which a contractor is liable if the architect's specification allows the contractor to choose the precise type of material within limits. It is established that, unless there is a special clause in the contract which provides to the contrary, the architect is liable for the suitability of the materials specified. If the architect specifies in such a way that the contractor has a choice, it may be assumed that the architect believes that no further restrictions are needed. Provided the material chosen was good of its kind, and it did not have actual knowledge of its likely bad effects, the contractor will not be liable.[33]

Employer's representative

This provision occurs only in SBC. Clause 3.3 allows the employer to appoint a representative who can act under the contract and exercise all the functions which the contract allows the employer to perform, or prescribes that the employer shall do; in other words, the employer's powers and duties. (The employer's agent under DB is quite different.)

In order to achieve this, the employer must issue written notice to the contractor, stating that the employer wishes the representative to deal with all the functions of the employer under the contract or whether any exceptions are required. Any exceptions must be set out in the notice. For example, the employer may not want the employer's representative to make certain decisions.

An intriguing footnote (39) suggests that, to avoid possible confusion over the quite distinct roles of the architect and the quantity surveyor and the role of the employer's representative, the employer's representative should not be either architect or quantity surveyor. Presumably, although it is not expressly stated, the employer's representative will be in reality a project manager in one of its several confusing manifestations.

Ownership of goods

Much difficulty has been caused by what is known as 'retention of title'. The position can be quite complicated. Clauses 2.24 and 2.25 of SBC, 2.17 and 2.18 of IC/ICD and clause 2.21 and 2.22 of DB are intended to provide that materials and goods stored on or off site, for which the employer has paid, become the employer's property.

In essence the problem is that the terms of the main contract bind only the parties to that contract; they cannot, subject to the Contracts (Rights of Third Parties) Act 1999, affect the rights of third parties such as suppliers. The supplier of the goods may have a retention of title clause in its contract of sale to the contractor which stipulates that the goods, in fact, remain the supplier's property until the supplier has been paid by the contractor. Therefore, if the employer pays the contractor for goods stored on or off site and the contractor does not pay the supplier, ownership of the goods will not pass to the employer. This is because the contractor cannot pass ownership until it has ownership itself. The goods still belong to the supplier, which may, in the event of the contractor's liquidation, take them away.

In such a case, the employer may be faced with the prospect of paying twice for the same goods; hence the traditional reluctance to certify or make payment for off-site materials. (See Chapter 5 for the provisions for off-site materials and goods.) There is an obligation to certify or pay for materials stored on site, but in either case the contractor may have to produce evidence of ownership or satisfy the architect in some other way that the employer will become the owner of the goods when the contractor has been paid.

A supplier's retention of title clause will normally be defeated once the goods are incorporated into the fabric of the building.[34]

That is a very simple exposition of a very complex problem, and contractors who encounter difficulties in this area should seek good legal advice. It tends, however, to be more of a problem for employers and suppliers.

It should be noted that all the contracts now contain a clause making it clear that no rights are conferred on third parties other than, in the case of SBC and DB, certain rights for purchasers, tenants and funders.

References

1 Bacal (Construction) Midlands Ltd v Northampton Development Corporation (1975) 8 BLR 88; Update Construction Pty Ltd v Rozelle Child Care Centre Ltd (1992) BLM vol. 9.2.
2 James Miller & Partners v Whitworth Street Estates Ltd [1970] 1 All ER 796.
3 Tameside MBC v Barlows Securities Group Services Ltd [1999] BLR 113.
4 Whittal Builders Co. Ltd v Chester le Street DC (1987) 11 Con LR 40.
5 The Brabant [1966] 1 All ER 961.
6 English Industrial Estates Corporation v George Wimpey & Co. Ltd (1972) 7 BLR 122.
7 McMaster University v Wilcher Construction Ltd (1971) 22 DLR (3d) 9.
8 Auto Concrete Curb Ltd v South Nation River Conservation Authority and Others (1994) 10 Const LJ 39.
9 London Borough of Merton v Stanley Hugh Leach (1985) 32 BLR 51.
10 C. G. A. Brown v Carr and Another [2006] EWCA Civ 785.
11 Liverpool City Council v Irwin [1976] 2 All ER 39.
12 Lawson v Wallasey Local Board (1982) 11 QBD 229.
13 John Mowlem & Co. Ltd v British Insulated Callenders Pension Trust Ltd (1977) 3 Con LR 64.
14 Greater London Council v Cleveland Bridge & Engineering Co. Ltd (1986) 8 Con LR 30.
15 Ibid.
16 West Faulkner Associates v London Borough of Newham (1995) 11 Const LJ 157.
17 John Barker Ltd v London Portman Hotels Ltd (1996) 12 Const LJ 277; Balfour

Beatty Construction Ltd v Mayor and Burgesses of the London Borough of Lambeth [2002] BLR 288.

18 Glenlion Construction Ltd v Guinness Trust (1987) 39 BLR 89.

19 National Coal Board v William Neal & Son (1984) 26 BLR 81.

20 Crown Estates Commissioners v John Mowlem & Co. Ltd (1994) 10 Const LJ 311.

21 Matthew Hall Ortech Ltd v Tarmac Roadstone Ltd (1997) 87 BLR 96.

22 University Court of the University of Glasgow v William Whitfield and John Laing (Construction) Ltd (1988) 42 BLR 66.

23 Brunswick Construction v Nowlan (1974) 21 BLR 27.

24 Edward Lindenberg v Joe Canning and Jerome Construction Ltd (1992) 62 BLR 147.

25 Plant Construction plc v Clive Adams Associates and JMH Construction Services Ltd [2000] BLR 137.

26 Co-operative Insurance Society Ltd v Henry Boot Scotland Ltd 1 July 2002 unreported.

27 Viking Grain Storage Ltd v T. H. White Installations Ltd (1985) 3 Con LR 52.

28 Stovin-Bradford v Volpoint Properties Ltd [1971] 3 All ER 570.

29 Davis Contractors Ltd v Fareham UDC [1956] 2 All ER 145.

30 Oldschool and Another v Gleeson (Contractors) Ltd and Others (1976) 4 BLR 103.

31 Holland & Hannen and Cubitts (Northern) Ltd v Welsh Health Technical Services Organisation (1981) 18 BLR 80.

32 Simplex Concrete Piles Ltd v Borough of St Pancras (1958) 14 BLR 80.

33 Rotherham Metropolitan Borough Council v Frank Haslam & Co. Ltd and M. J. Gleeson (Northern) Ltd (1996) EGCS 59.

34 Reynolds v Ashby [1904] AC 406.

2 Insurance

2.1 General

Insurance is a highly specialised field. Although the details of policies are best left to the experts, it is essential that the employer and the contractor know their respective rights and obligations under the contractual provisions. SBC and DB deal with indemnities and insurance in clause 6 and schedule 3. IC and ICD deal with them in clause 6 and schedule 1, and MW and MWD deal with them in clause 5 in each case. They can be broken down into the following parts:

- indemnities – injury to persons and property;
- insurance – injury to persons and property;
- insurance – liability of the employer (not MW or MWD);
- insurance of the Works – new Works – existing structures;
- remedies if a party fails to insure.

SBC, DB, IC and ICD provisions are virtually identical. MW and MWD provisions are less detailed, with some significant omissions, and they will be considered later.

2.2 Injury to persons and property

There are two important definitions in clause 1.1 of SBC, DB, IC and ICD. Although there are no equivalent definitions in MW or MWD, the wording of the relevant clauses amounts to much the same. 'Employer's persons' are persons engaged or authorised by the employer, but specifically excluding the contractor and the contractor's persons, statutory undertakers, architect (not relevant to DB) and quantity surveyor. 'Contractor's persons' are the contractor's employees, agents and any persons employed on the site except the architect, quantity surveyor, employer, employer's persons and statutory undertakers.

The contractor indemnifies the employer and takes liability in the case of any loss, expense, claim or proceedings as a result of carrying out the Works in respect of:

- personal injury or death of any person unless and to the extent that it is due to the act or neglect of the employer or any employer's persons;
- injury or damage to property of all kinds except the Works and existing structures insured by the employer which and to the extent that it is due to the negligence or default of the contractor's persons.

The contractor is also liable if the injury to persons or property is a result of breach of statutory duty and, in addition to its servants or agents, it is made liable for any contractor's persons.

Thus, if a person is injured owing to the Works, the contractor is liable unless it is the employer's fault. If property is damaged, the contractor is liable only if it is the contractor's fault.[1] It is clear that the contractor is still obliged to indemnify the employer against claims for personal injury or death even if the employer's neglect is partially responsible. In such a case, of course, the contractor's liability would be reduced accordingly. A similar formula has been introduced to maintain the indemnity to the employer even if the contractor's negligence is only a part cause of injury to property.

In practice, a person suffering injury to person or property would usually claim against the employer, who would, by virtue of this clause, join the contractor as a third party in any proceedings. It should be noted that the contractor is not liable under this clause for any loss or damage to the Works unless they have been taken into partial possession or a section completion certificate has been issued under the relevant provisions of SBC, DB, IC or ICD.

It may be thought superfluous to have an indemnity clause when the following clause requires the contractor to take out insurance against just the same liabilities. But the indemnity is important. If a claim were successful against the employer and the insurance company refused to meet the claim for some reason, the contractor would retain liability. The level of insurance required is to be inserted in the Contract Particulars.

It should be noted that the contractor does not provide the employer with indemnity against the results of the employer's own negligence. To provide such indemnity, the clause must be expressed in very clear words.[2]

2.3 Liability of employer

SBC, DB, IC and ICD clause 6.5 are in identical terms intended to cover the liability of the employer if damage is caused to any property other than the Works by the carrying out of the Works. Such occurrences as collapse, subsidence, heave, vibration, weakening or removal of support or lowering of ground water arising from the carrying out of the Works are included. This clause operates when the damage which occurs is not due to negligence on the part of either employer or contractor or their servants or agents. For example, the contract may call for piling operations on a town-centre site and, despite careful design and conscientious work, cracking occurs in adjacent buildings. Because the contractor is liable only for such damage caused by its negligence, without this clause the employer would have to foot the bill.[3] The clause envisages that a sum may be included in the contract documents representing the indemnity required and the architect may instruct the contractor to take out appropriate insurance. The insurance must be taken out by the contractor in joint names with insurers approved by the employer, with whom policy and premium receipts must be deposited.

The contractor is expressly stated to have no liability in respect of injury or damage to any person, the Works, site or any property caused by ionising radiation or contamination by radioactivity from any nuclear fuel or from any nuclear waste from the combustion of nuclear fuels, radioactive toxic explosive or other hazardous properties of any explosive nuclear assembly or component, or by pressure waves caused by aircraft or other aerial devices travelling at sonic or supersonic speeds. These risks are called 'excepted risks'. Any such damage, therefore, will be the sole responsibility of the employer.

2.4 Insurance of the Works

There are two types of insurance risks in the contract:

- '*Specified perils*' *insurance*. This is insurance which provides for insurance against fire, lightning, explosion, etc.
- '*All risks*' *insurance*. This is insurance against physical loss or damage to work executed and site materials, but *excluding* the cost of repairing, replacing or rectifying property which is defective owing to wear and tear, obsolescence, deterioration, rust or mildew or any work executed or materials lost or damaged as a result of its own defect in design, plan, specification, materials or workmanship or any other work executed which is lost or damaged if such work relied for its support or stability on the defective work.

Other exclusions include loss or damage arising from war, nationalisation or order of any government or local authority, disappearance or short-age revealed only on the making of an inventory and not traceable to any identifiable event, the exceptions with regard to ionising radiation, etc. Therefore, risks such as impact, subsidence, theft and vandalism are included in this type of insurance. There used to be certain other exclusions applicable only to Northern Ireland (civil commotion, unlawful and malicious acts, etc.). These are now introduced into contracts used in Northern Ireland, together with other amendments, by means of an adaptation schedule obtainable from the Royal Society of Ulster Architects.

Insurance of the Works falls into two categories: new Works and work on existing structures. There are three standard provisions: SBC and DB schedule 3, options A, B and C (IC/ICD schedule 1, options A, B and C). The first two clauses in each contract relate to new work where either the contractor or the employer, respectively, has the responsibility to insure. The type of insurance to be taken out is 'all risks'.

Not only the Works, but the value of any unfixed materials or goods delivered to the site must be insured, including a percentage stated in the Contract Particulars to cover professional fees. Full reinstatement value is required, which will, of course, be greater than the simple value of everything on the site, but it does not include the additional costs of carrying out subsequent work later than intended or the loss suffered by the employer due to delay. The insurance must be taken out in joint names with insurers approved by the employer, with whom policy and premiums must be deposited. The contractor should obtain good advice. The risk does not include consequential loss.[4]

The insurance must be maintained up to and including the date of the issue of the practical completion certificate. This will almost always be some days after the actual date of practical completion and prevents the premature termination of insurance cover if, for some reason, expected practical completion does not take place. Provision is made for the cover to cease if termination of the contractor's employment occurs, and this is the case even if either party contests the termination.

The insurance is to be taken out in the joint names of the employer and the contractor. The effect of this is that the insurers now have no right of subrogation against one or other of the parties if damage occurs due to the other party. Subrogation, in this context, is the right of the insurer to stand in the shoes of the insured party and recover, from the party causing the damage, amounts paid out. Since both parties are insured on a single policy, if one of the 'all risks' occurred owing to the employer's negligence, while it is the contractor's obligation to insure in joint names, the position would be covered by the policy and the insurer would not be able to claim against the employer. Where the contractor wishes to use its existing annual policy for option A, it must contain terms which equate it to a joint names policy. The employer's interest must be endorsed and the annual renewal date must be stated in the Contract Particulars. Option B, where the employer is to insure, is little used in practice.

Financial responsibility

If there is any element of under-insurance or any excess payable on the insurance, this is the responsibility of the party that has the obligation to insure. Where the contractor has the obligation to insure, insurance money is to be payable to the employer, who must issue, through the architect, special reinstatement certificates on the same date as normal interim certificates as repair work proceeds until the insurance money is fully certified less only the amount of professional fees. The contractor is entitled to no money additional to insurance money in respect of the insured matter, but it is entitled to interim certificates in the normal way for other work carried out even though some of that work has been damaged by an event which resulted in the insurance payment to the employer. Thus, except for the possibility of under-insurance or excess, neither the contractor nor the employer should suffer any financial loss. If the employer insures, the remedial work is to be treated as if it were a variation and the contractor is to be paid its full cost.

If either the contractor or the employer insures, the contractor must immediately give written notice of any damage to employer and architect. The contractor must state its nature, extent and location. After inspection by the insurers, if required, the contractor must proceed with restoration and removal of debris.

Existing structures

If the work consists of alterations or extensions to existing structures, the employer must insure the existing structure and contents, the Works, and all unfixed materials delivered to the site and intended for incorporation. The insurance is in two parts:

- specified perils insurance for the existing building and contents;
- all risks insurance for the new work.

Both parts are to be taken out in joint names. The distinction between types of insurance is important. Where, for example, the contractor or its sub-contractor causes damage to a water pipe in an existing building, the contractor will be liable for the resultant damage caused by the escaping water, which will not fall under the descriptions of flood, escape of water from water tanks, apparatus or pipes as described in 'specified perils'.[5]

There has been much uncertainty surrounding the position when the employer is to insure the Works and existing structures. There have been decisions in the courts to the effect that even if the contractor by its negligence caused fire damage to the Works or existing structures which were insured by the employer, the employer had to bear the risk.[6] The decisions of the courts, combined with changes in the wording of the insurance clauses, have been confusing. In one instance, what appeared to be a sound judgment excluding the situations where the contractor was negligent has been overturned by the Court of Appeal,[7] which seems to suggest that the courts themselves are not finding it easy to interpret the clauses. In the light of the different decisions, the basic principle under the 2005 JCT contracts appears to be that if damage which falls under one of the categories of specified perils is caused to the existing structure, a claim can be made under the employer's insurance even where the damage is a result of the contractor's negligence.

In the event of loss or damage, the contractor must give immediate written notice to the employer and the architect. At this point, either party may terminate the contractor's employment if it is just and equitable so to do (see section 7.2). If neither party opts for termination, or the notice is decided against by an arbitrator, the contractor must make good the loss and damage, remove debris and proceed with carrying out the Works as before. The contractor's work is to be treated as a variation and the contractor is to be paid accordingly.

Restoration

Note that the contractor's obligation to commence restoration under all the SBC and DB schedule 3 (IC/ICD schedule 1) insurances begins as soon as the insurers have carried out any inspections they deem necessary and not, as previously, when the claim is accepted. The contractor is, therefore, put into the position of carrying out rectification work before it knows whether any insurance money will be paid out.

Failure to insure

In regard to any of the insurance which is the contractor's responsibility, it is required to produce evidence of insurance on demand to the employer unless it has already deposited the policy and premium receipts. If the contractor fails to insure, the employer may take out and maintain the appropriate insurance, deducting the cost from monies due or to become due to the contractor. Alternatively, the employer may sue for the debt.

The contractor has similar powers if the employer fails to insure new work where that clause applied. If the employer fails to provide evidence of insurance, the contractor may itself take out the necessary insurance and is entitled to have the amount of the premium added to the contract sum on production of a receipt. Where the employer is required to insure existing structures and extensions, etc., the contractor has unusual powers if the employer fails in this duty. The contractor has the right to enter and inspect the existing structure for the purpose of making an inventory and survey. Where a premium has been paid to take out the insurance, the contractor is again entitled to have its cost added to the contract sum. The only difference between private and local authority use is that the local authority is not obliged to produce evidence of insurance on demand and the contractor has no power to inspect or to take out insurance if the local authority fails to do so. These provisions, in any case, apply only if option B or C (in SBC, DB, IC or ICD) is used.

Liability to insure

The contractor's liability to insure ends at the date of termination of the contractor's employment or the issue of the practical completion certificate, or on partial possession as regards the relevant part, but it is important that the contractor obtains formal certification before dispensing with such insurance.[8]

Insurance of the contractor's plant and tools is not provided for under either contract and is the contractor's liability. Most contractors do carry this sort of insurance.

Under clause 2.6 (SBC, IC and ICD) or clause 2.5 of DB, if the employer wishes to use the Works to store goods, the employer or the contractor must obtain the insurer's agreement and pay any additional premium which may be required. It should be noted that early use of the Works under these clauses does not rank as partial possession, and if the contractor is in culpable delay, liquidated damages for the part in early use are still recoverable.[9]

If employer does not wish to insure

If the employer does not wish to use options A, B or C, model clauses have been prepared which enable the employer to take either the risk or the sole risk for damage caused by any of the 'all risks' without the need to insure. Contractors should note, however, that if the employer opts to accept the risk (rather than the sole risk), the employer has the right to refuse payment to the contractor for rectification work to the extent that the loss or damage is due to the negligence of the contractor. If the employer opts for taking the risk only, the contractor should consider whether it should itself insure against possible liability.

Terrorism cover

Clause 6.10 of SBC, DB, IC and ICD provides that if either employer or contractor is notified by the insurers that terrorism cover will cease from a particular date, either party as appropriate must notify the other. The employer has just two options: either to require in writing that the Works continue to be carried out, or to specify that the contractor's employment will terminate on a date after the date of the insurer's notification but before the date of cessation of cover. If the option is termination, clauses 8.12.2–8.12.5 in each of the four contracts apply (excluding clause 8.12.3.5). If the decision is not to terminate and damage is suffered as a result of terrorism, the resulting work is treated as a variation.

Joint Fire Code

Clauses 6.13–6.16 of SBC and DB and clauses 6.11–6.14 of IC/ICD apply where the Contract Particulars state that the Joint Fire Code applies. The Joint Fire Code means the Joint Code of Practice on the Protection from Fire of Construction Sites and Buildings Undergoing Renovation (published by the Construction Confederation and the Loss Prevention Council). This should be the case if the insurer of the Works requires the employer and the contractor to comply with the Code. These clauses place an obligation on both parties to comply with the Code, and each indemnifies the other for the consequences of any breach of the Code. There are also time stipulations within which the contractor must carry out remedial measures to rectify a breach. Where the insurer has stated that the Works are a 'Large Project', the Code specifies additional requirements.

2.5 Professional indemnity insurance

DB and SBC clauses 6.11 and 6.12 and ICD clauses 6.15 and 6.16 require the contractor to take out and maintain professional indemnity insurance in respect of the contractor's design. The type and limit of indemnity are to be stated in the Contract Particulars. The period for which the insurance must run from practical completion of the Works is also to be stated in the Contract Particulars. Usually this will be six years if the contract is executed under hand and twelve years if it is a deed. There is a saving provision which makes the contractor's obligation dependent on insurance being available at commercially reasonable rates. If the insurance becomes unavailable at such rates, the contractor must give notice to the employer, and the employer and the contractor must see how they can best protect their positions without such insurance. If there was an easy answer to that question, no doubt the contract would have already included it as a consequential clause.

This is a welcome improvement to the contracts. If there is a design fault, the employer can be expected to claim against the contractor and it is sensible for the contractor to have insurance to cover such an eventuality. It also protects the employer against the difficulty which can arise when the employer has a good claim but the contractor is clearly not in a position to meet it.

2.6 Sub-contractors

Protection for sub-contractors is provided under SBC, DB, IC and ICD clause 6.9 whereby the contractor or employer, as appropriate, will ensure broadly that the joint names policy either:

- provides for recognition of each sub-contractor as an insured under the joint policy; or

- includes a waiver by the insurers of the right of subrogation against any sub-contractor.

Protection is offered to sub-contractors in respect of loss or damage to the Works or to site materials by specified perils.

2.7 MW and MWD insurance

The insurance position under MW and MWD is broadly similar to that under SBC, DB, IC and ICD, but it is much more briefly set out and there are differences, mainly omissions.

The contractor has the same indemnity and insurance liabilities in respect of injury or damage to persons or property, but there is no provision for insurance to cover damage to adjacent property if no one is negligent. Presumably this is because the contract is primarily intended for small Works. The size of the job, however, is no indication of the possibility of such damage, and each project should be assessed on its merits and a suitable clause inserted if necessary. It should also be noted that there is no provision for professional indemnity insurance in MWD.

Essentially, clause 5.2 makes the contractor liable for damage to property, except the Works or the existing property if insured by the employer under clause 5.4B, to the extent that the damage is due to the contractor's or any sub-contractor's negligence or default. The contractor will be liable to a partial extent if it is partially at fault and damage is caused to surrounding buildings or passing vehicles. The Works are expressly excluded, but it is the contractor's obligation to carry out and complete the Works in accordance with the contract documents. Therefore, the contractor has to properly complete the Works in any event. In other words, it has to redo the work if it is defective. It would be a nonsense if the contractor could claim on the employer's insurance for failure to carry out the Works in accordance with the contract. The contractor is obliged to carry appropriate insurance.

If the Works are alterations or extensions of an existing structure, the employer is obliged to insure the Works and the existing structure under the provisions of clause 5.4B. Effectively, the employer and contractor agree that if there is damage caused by one of the specified perils, this clause will deal with the situation.[10] Under previous contract wording, damage by fire due to the contractor's negligence has been held not to be so covered.[11] It is suggested that the new clause 5.4B covers all specified perils even if due to the contractor's negligence.

There are only two possibilities in respect of damage to the Works: insurance of new Works by the contractor or insurance of existing structures by the employer. The risks to be insured are identical to specified perils in SBC, and the parties should consider whether, in any particular case, more extensive insurance is required. There is no provision for termination after loss or damage if it is just and equitable, and the parties must make their own arrangements in such circumstances. Otherwise, the position after damage is the same as under the other contracts. The party responsible for insuring is required to produce evidence of insurance to the other on request but, if the other defaults, there is no provision for either the employer or the contractor to insure, nor can the cost of such premiums be recovered. Since the situation is neither provided for in the termination clauses nor, it is thought, sufficient to allow repudiation at common law, it might be best for the party not in default to take out the appropriate insurance itself, after due notice, and then take the matter to adjudication or arbitration to claim damages for breach of contract. Hopefully this will be a rare occurrence.

It is essential that the contract provisions are read carefully to ensure that the appropriate insurances are in force before the contractor takes possession of the site. Insurances should not simply be passed to the employer without any kind of comment. It is the architect's, or in the case of DB the employer's agent's, duty either to check the contractor's insurance, to ask an expert to do so or to make sure that the employer obtains expert advice.[12] The JCT produced a detailed guide to the insurance clauses in previous editions of JCT contracts (Practice Note 22). Although not completely applicable to current insurance clauses, the Practice Note contains information which is still useful. Architects, contractors and sub-contractors are advised to obtain copies and give them careful study.

References

1 City of Manchester v Fram Gerrard Ltd (1974) 6 BLR 70.
2 AMF International Ltd v Magnet Bowling Ltd [1968] 2 All ER 789.
3 Gold v Patman & Fotheringham Ltd [1958] 2 All ER 497.
4 Kruger Tissue Industrial Ltd v Frank Galliers Ltd and DMC Industrial Roofing & Cladding Services and H&H Construction (1998) 51 Con LR 1; Horbury Building Systems Ltd v Hampden Insurance NV [2004] EWCA Civ 418.
5 Computer & Systems Engineering plc v John Lelliot (Ilford) Ltd (1989) *The Times* 23 May 1989.
6 Scottish Special Housing Association v Wimpey Construction UK Ltd [1986] 3 All ER 957.
7 Scottish & Newcastle plc v GD Construction (St Albans) Ltd [2003] EWCA Civ 16.
8 English Industrial Estates Corporation v George Wimpey & Co. Ltd (1972) 7 BLR 122.
9 Skanska Construction UK Ltd v Egger (Barony) Ltd [2004] EWHC 1748 TCC.
10 National Trust for Places of Historic Interest or Natural Beauty v Haden Young (1994) 72 BLR 1.
11 London Borough of Barking and Dagenham v Stamford Asphalt Co. Ltd (1997) 82 BLR 25.
12 Pozzolanic Lytag Ltd v Bryan Hobson Associates (1998) CILL 1450.

3 Third parties

3.1 Assignment and sub-letting

Under SBC, IC, ICD, MW, MWD and DB, the contractor has no automatic right to sub-let work (SBC clauses 3.7–3.9, IC/ICD clauses 3.5 and 3.6, MW/MWD clause 3.3 and DB clauses 3.3 and 3.4).

Assignment

It is refreshing to see that assignment is no longer coupled with sub-letting in any of the contracts. It is a different concept. Assignment is dealt with in clause 7.1 of SBC, DB, IC and ICD and in clause 3.1 of MW and MWD.

Assignment occurs when a right or duty is legally transferred from one party to another. After the transfer, the original party retains no interest in the right or duty transferred. When this is carried out properly, it is termed novation and involves the formation of a new contract.

The parties to a building contract are, of course, the employer and the contractor. In broad terms their rights are to receive a completed building and to receive payment respectively. Their duties echo the rights. The employer's duty is to pay and the contractor's duty is to complete the work. Under the general law a party is entitled to assign the benefit of its rights to a third party, but neither employer nor contractor may assign their duties. An express term in the contract may modify this position. Thus, in the absence of any express term, a contractor may assign its right to payment to a third party in return for money 'up front' in order to enable it to carry out the contract. An employer may assign the right to the completed building to another for payment.

All the three contracts under consideration contain an express term forbidding assignment of rights or duties by either party unless both parties agree. There is no stipulation that a refusal of consent must be reasonable. There appears to be no good reason why the contracts should not be amended to permit assignment of rights and, if a contractor is anxious to raise working capital in this way, it should be brought to the attention of the employer at tender stage. If a contractor or employer attempts to assign rights despite the clause to the contrary, the assignment will be ineffective.[1] The clause will also prevent assignment of the right to damages for breach of contract. However, if the purported assignee cannot sue the contractor, because the assignment is prohibited, the employer may be able to recover damages (presumably on behalf of the purchaser) if it could be seen by the contractor that a subsequent purchaser of the building might suffer loss. This is a difficult point in practice.[2]

Clause 7.2 of SBC and DB applies only when so stated in the Contract Particulars. It allows the right to bring proceedings against the contractor in the name of the employer to be assigned to enforce any term of the contract made for the benefit of the employer. The power may be used if the employer sells or leases the whole of the premises comprising the Works after practical completion. Without this important clause, a future purchaser or lessee of the property has very little chance of obtaining satisfaction through tortious remedies.[3] It is becoming very common now for the contractor, sub-contractors, suppliers and all the consultants to enter into warranties which themselves contain assignment rights and often very much more onerous conditions than the standard form of contract. The third party is not entitled to dispute enforceable agreements made between the employer and the contractor before the date of the assignment. See section 3.5 for details of third party rights and warranties.

Sub-letting

Sub-letting is quite different from assignment. It is the delegation rather than the transfer of a duty. For example, if a contractor sub-lets plumbing work to a sub-contractor, the main contractor remains responsible if the sub-contractor fails to perform or performs badly. If, however, the contractor were able by novation to assign its duty to carry out the contract to a second contractor, it would be the second contractor that would be liable if the work were done badly.

Perhaps in an attempt to clarify the position, clauses 3.7.1 of SBC, 3.3.1 of DB, 3.5 of IC/ICD and 3.3.1 of MW/MWD provide that even though part of the Works may be sub-let, the contractor remains wholly responsible for carrying out and completing the Works in all respects. A similar provision is included in clause 2.2 of DB and schedule 2 paragraph 13 of IC/ICD in respect of named sub-contractors.

The right to sub-let work is now similar in SBC and DB, but there are subtle differences in IC/ICD and MW/MWD. There is a complete prohibition on sub-letting without consent in SBC, DB, MW and MWD. Under IC and ICD there is a prohibition on sub-letting other than with the conditions in clauses 3.6 or 3.7 applied unless the architect's consent has been obtained. It could be argued that, under IC and ICD, the contractor can sub-let without consent provided the conditions in clauses 3.6 or 3.7 are applied. That may not be what the contract draftsman had in mind.

Where the architect's written consent is required, there is a proviso that the architect's consent must not be withheld unreasonably. A similar condition relating to the consent of the employer to sub-letting work (3.3.1) and design (3.3.2) is contained in DB. Note that the contractor is obliged to obtain consent only to the fact of sub-letting, not to the sub-contractors themselves. In practice, of course, the architect may well consider that it is reasonable to refuse consent unless the contractor reveals the names of the sub-contractors.

All the contracts now require the contractor to ensure that sub-contracts include a term to entitle the sub-contractor to simple interest at 5 per cent above Bank of England base rate if the contractor fails to pay money due by the final date for payment. In addition, if the contractor's employment is terminated, for any reason, under the main contract, the employment of the sub-contractor under the sub-contract is to be similarly terminated. This provision, of course, will be effective only if it is included in the sub-contract, because the sub-contractor is not a party to the main contract and cannot be bound by its terms. Contractors should, therefore, take care to use the forms of sub-contract devised for use with the standard main contract forms or ensure that all matters in the main contract which might affect the sub-contract are incorporated in the sub-contract. SBCSub/C and SBCSub/D/C are the domestic forms of sub-contract for use with SBC, ICSub/C is the equivalent for use with IC, ICSub/D/C with ICD and DBSub/C is used with DB.

MW and MWD have nothing more to say about either assignment or sub-contracting and there is no standard form of sub-contract. SBC, IC, ICD and DB contain further provisions. Each sub-contract must include a provision granting access for the architect to workshops and sub-contractors' premises and providing for the execution and delivery of sub-contract warranties within fourteen days of the contractor's written request, if applicable.

Ownership

An attempt is made in SBC clause 3.9, IC/ICD clause 3.6 and DB clause 3.4 to protect the employer against retention of title clauses preventing ownership of materials passing to the employer. The dangers were highlighted in a case concerning a sub-contractor.[4] This question is discussed in section 1.4. The clauses stipulate that the contractor must include conditions with regard to unfixed materials provided by the sub-contractor. The conditions are set out in the clauses. In essence they are as follows:

- Unfixed goods must not be removed from the Works except with the architect's consent.
- If the value of the goods has been included in a certificate which the employer has paid to the contractor, the goods become the property of the employer and the sub-contractor will not deny it.

- The goods will be the property of the contractor if it has paid the sub-contractor before the value has been certified.

If for any reason the contractor does not include such conditions, it is in breach of contract. That is likely to be small comfort to the employer, because problems with retention of title are likely to occur only if the main contractor becomes insolvent. In such a case, suing for breach of contract would seem to be a fruitless exercise. The provision may, however, lead to architects asking to inspect sub-contracts to ensure that the conditions have been incorporated. Even if such provisions are included in sub-contracts, they will not be effective against sub-sub-contractors or suppliers to the sub-contractors who may have a retention of title clause in their sub-sub-contracts or contracts of sale. The only way for the employer to be sure is if the provisions are stepped down to the earliest point of supply or if an amendment is made so that the architect is not required to certify (and, under DB, the contractor cannot apply for payment) for unfixed materials on site. Although one form of contract contains provisions for putting that option into effect, it is unlikely to be popular with contractors, and rightly so.[5]

IC and ICD go on to deal with 'named sub-contractors', and DB, in the supplemental provisions, deals with persons named as sub-contractors in the Employer's Requirements, a provision that will be discussed at length in section 3.2. Nominated sub-contractors and suppliers have now been removed from SBC.

Domestic sub-contractors

SBC contains further important provisions with regard to what it terms 'domestic sub-contractors'. This is now the only kind of sub-contractor in SBC, since nomination has been removed. Clause 3.8 provides the employer with a way of narrowing the choice of sub-contractor in appropriate cases. The system is that the work to be done must be measured or described adequately in some other way in the bills of quantities so that it can be priced by the contractor. A list of persons or firms is provided in the bills from which the contractor must choose to carry out the work. The contractor has sole discretion in selecting the firm.

The list must contain at least three names. Presumably the architect will have contacted all the firms on the list to make sure that they are willing and able to carry out the work. Either the employer (or the architect acting on behalf of the employer) or the contractor may add more names to the list at any time before a binding sub-contract is entered into in respect of the particular work. The only proviso is that the other party must consent to the additional names. Consent, however, is not to be unreasonably withheld. This provision can be useful to a contractor who wishes to use a firm not included on the original list. It seems that the only reasonable ground for refusing consent, so far as the employer is concerned, would be that the suggested additional firm is not capable of carrying out the work to the standard required by the contract.

Note that the additional names may be added by either party even after the main contract is let. This gives maximum flexibility to the contractor to take advantage of competitive prices. Effectively, irrespective of the sub-contractors named in the contract, the contractor can use any sub-contractor it wishes simply by adding the name to the list – subject of course to the employer being unable to raise a reasonable objection.

If at any time before the contractor has entered into a binding sub-contract the number of firms able and willing to carry out the work falls below three:

- The employer and the contractor must agree on the addition of more firms so that the list comprises at least three.
- Alternatively the contractor may carry out the work itself, and in so doing, it may sub-let the work to any sub-contractor of its choice provided the architect gives consent.

Although the contract requires the parties to agree the additional names, as a matter of practicality and law an agreement to agree has no validity. No doubt the parties will try to agree, failing which, the contractor appears to have free rein to employ any sub-contractor of choice.

A sub-contractor chosen from the list by the contractor becomes a domestic sub-contractor. The employer will not, thereafter, be concerned with any problems of delay, financial claims or termination of employment. These are solely the concern of the contractor and the sub-contractor involved. The contractor, of course, may be able to found a claim on events relating to the sub-contract work as if the work was carried out by the contractor's own operatives (for example, extension of time for exceptionally adverse weather conditions). The contractor, however, remains responsible for the work of its domestic sub-contractors, and for any defects therein, to the employer.

Sub-contract

This is not the place to discuss the forms of sub-contract, but there are some general points which deserve mention if only to emphasise the importance of having a sub-contract which sets out the rights and duties of the parties in a realistic manner.

Many problems between main and sub-contractors arise because a project is not ready for a sub-contractor to commence work on the date anticipated, or the sub-contractor is prevented from maintaining reasonably expected progress owing to delays for other reasons. If the commencement of the sub-contract is delayed, the sub-contractor may say that it wants more money, or even that it cannot, at a later date, fit the work into its programme. Whether the sub-contractor can lawfully take that stance depends upon the terms of the sub-contract. If there are no express terms or agreement about the date for commencement, the law will not imply that the contractor must have the site ready for the sub-contractor by any particular date. The most which will be implied is that the contractor must have the site ready within a reasonable time.[6]

Similarly, in the absence of an express term of the sub-contract, there will be no implied term that the contractor must ensure that there is sufficient work available to enable the sub-contractor to maintain reasonable progress and execute the sub-contract Works in an efficient and economic manner.[7] However, it is now established that sub-contract form DOM/1 simply requires the sub-contractor to finish its work by the date fixed in the sub-contract. Provided it does that, it may plan and perform the work as it pleases.[8] Although DOM/1 has now been replaced by SBCSub/C, it is thought that the principle also applies to the new sub-contract form and also to the new sub-contract forms for use with the other main contracts under consideration here. Thus, if the sub-contract contains a term requiring the sub-contractor to proceed with the sub-contract works at such times and in such manner as the contractor directs, the sub-contractor will be required to do exactly as stated. In many instances the contractor will supply a sub-contractor with a programme, but it will seldom be a contract document. Only if it is a contract document will the sub-contractor be obliged to work to it and an instruction to work differently would constitute a variation.[9]

In the absence of a warranty, employers will find it very difficult to take direct action against sub-contractors in tort. The only exceptions are likely to be if the sub-contractor's negligence causes damage to property other than the Works or if there is a danger or imminent danger to health or safety. However, where there is a warranty, its terms will not necessarily constitute the whole of the sub-contractor's liability, and the courts are likely to find a concurrent liability in tort.[10] This liability may sometimes exceed the sub-contractor's liability in contract.[11]

3.2 Named sub-contractors

IC and ICD make provision for something called naming. This is not just a matter of terminology. The provision is vastly different from the former nomination procedures in JCT 98. DB also has a naming procedure in the supplemental provisions. The idea is to provide the employer with a means of ensuring that particular parcels of work are carried out by sub-contractors of the employer's choice. There is always a conflict here between choice by one party and responsibility for that choice being put on another party. The courts have never been greatly in favour of the system, as previous decisions show all too clearly.

Clause 3.7 provides for persons to be 'named', and the detailed provisions are now relegated to schedule 2. There are two situations:

- where work is included in the contract documents and priced by the contractor to be carried out by a person named in the documents;
- where there is a provisional sum and the architect issues an instruction naming a person to carry out the work it represents.

In the first situation the contractor must enter into a sub-contract (ICSub/ NAM/C) with the named person within twenty-one days of entering into the main contract. The main contract will become binding when the employer accepts the contractor's tender. There is, therefore, very little time. A substantial quantity of documentation is involved which is obviously intended to be part of the tender documentation and, thereafter, part of the contract documents. That presupposes that all sub-contract tenders will be received well before the main contract tenders are invited. This first option is not something which will appeal to an architect when time is pressing. The architect will be more likely to take the second option.

If the contractor cannot enter into a sub-contract in accordance with the particulars in the contract documents, it must notify the architect. An architect who is satisfied that the particulars have prevented the execution of the sub-contract has three options:

- alter the particulars, creating a variation;
- omit the work, creating a variation;
- omit the work and substitute a provisional sum for which an instruction is required.

Note that the architect can alter the particulars only in so far as they are not contract terms. Only the parties to a contract can agree to vary its terms. If the work is omitted in its entirety, the employer may use directly employed labour to carry it out under clause 2.7.

In the second situation, the instruction on a provisional sum can arise:

- after the contractor has notified the architect that the particulars are preventing execution of the sub-contract;
- if the provisional sum is substituted before the sub-contract is signed;
- if there is already a provisional sum in the contract documents.

The instruction must name a person, describe the work and include all the complex documentation (ICSub/NAM/IT and ICSub/NAM/T). In this instance, the contractor has fourteen days from the date of the instruction in which to make a reasonable objection. If a reasonable objection is made, the contract leaves the position open. Presumably the architect must name another firm and so on. The procedure could be protracted. Fortunately, the architect now has power to make an extension of time under the broad relevant event clause 2.20.6. If there is no objection, the contractor must enter into a sub-contract with the named person.

There is a form of employer/sub-contractor agreement which should be completed (ICSub/NAM/E). This provides the only redress for the employer if the sub-contractor fails in the design or selection of materials for which it has undertaken responsibility. The contractor has no liability in these areas. It should be noted that this is the case even under ICD where the contractor is given design responsibility in certain circumstances for CDP work. What is worse, from the employer's point of view, no other sub-contractor is so liable, since the employer's only redress in the case of ordinary (clause 3.5) sub-contractors is through the main contractor. In general, however, the contractor is responsible for the normal workmanship and materials aspects of a named person's work.

Termination

If the named person's employment is terminated, the architect may:

- name another person;
- instruct the contractor to make its own arrangements to do the work;
- omit the work still to be completed.

In the last case, the employer may employ direct labour to carry out the work.

The consequences of termination are complex, depending upon whether the work was originally included in the contract documents or arises as a result of an instruction regarding the expenditure of a provisional sum. Briefly, in the first case, if the architect has exercised the option of naming another person, the contractor is entitled to an extension of time, but not a claim for loss and/or expense. In all other circumstances the contractor is entitled to both time and money. It is important to note, however, that if the termination occurs through the contractor's fault, it can have none of these benefits.

The contractor must take whatever action is necessary within reason to recover from the named person any loss sustained by the employer as a result of the termination. The contractor need not take any legal proceedings unless the employer agrees to pay any legal costs incurred.

Named sub-contractor under DB

In order for the provisions contained in schedule 2 to bite, the Employer's Requirements must state that certain work is to be carried out by a named sub-contractor. The procedure is that the contractor must execute a sub-contract soon after entering into the main contract. If it cannot do so, the employer is entitled to amend the Requirements if that will make it possible or omit the work and issue instructions about it. If appropriate, the employer may simply instruct the contractor to do it itself or select a suitable sub-contractor, or the employer may simply engage direct operatives under clause 2.6. The employer's consent, which as usual may not be unreasonably withheld, is needed if the contractor wishes to terminate the sub-contractor's employment.

After termination, the contractor must finish the work itself. The contractor is obliged to use reasonable diligence to recover from the defaulting sub-contractor monies due to the employer. Otherwise, unless the termination is due to the contractor's own default, the finishing work is treated as a change (the name for a variation under this contract) for which, of course, the employer must pay. The contractor must include a term in the sub-contract designed to enable it to claim money even though the employer rather than the contractor has suffered the loss. Such a clause is thought to be effective.[12]

MW and MWD

It is possible, although not advisable, effectively to nominate sub-contractors in MW and MWD by:

- naming the firm in the contract documents;
- naming the firm in an instruction regarding a provisional sum (clause 3.7);
- including a specially worded nomination clause in the contract.

The big problem with the first two options is that there is no provision for the consequences, which case law has shown can be considerable. To include a special clause is probably the most satisfactory, but it is counter to the spirit of this simple form of contract, and is probably sufficient to class the whole contract as the employer's written standard terms of business for the purpose of the Unfair Contract Terms Act 1977.

Introducing nomination into these contracts, at a time when even the JCT appears to concede that it is best to omit it from SBC, seems to be a bad move.

Points

A final point in regard to sub-contractors. In general terms it can be said that the inclusion of a 'pay when paid' clause in the sub-contract will be effective against the sub-contractor, which has completed its work, if the employer becomes insolvent before paying to the main contractor sums due to the sub-contractors.[13] It should be noted that the Housing Grants, Construction and Regeneration Act 1996 section 113 states that payment provisions which attempt to make payment dependent on receipt of payment from a third party (pay-when-paid provisions) are ineffective except in the case of third party insolvency. Although the effect is reduced, from a contractor's point of view, it is still worth while having a 'pay when paid' clause to afford protection in the event of employer insolvency. Of course, it is relatively simple to devise other clauses which have a very similar effect to pay-when-paid clauses, but which are not caught by the Act.

3.3 Employer's directly employed contractors

When the contractor is in possession of the site it is said to have a licence to be there in order to carry out its contractual obligations. Ideally, those obligations should embrace every aspect of the Works and the construction should be entirely under its direction and control. The drawings and bills of quantities should be complete and accurate and the architect should be required only to make occasional visits to keep an eye on progress. Life, certainly a contractor's life, is rarely like that.

Quite apart from the contractor's own shortcomings, inaccuracies in drawings and a steady stream of variations, SBC, IC, ICD and DB provide for the possibility of other contractors, quite separately engaged by the employer, arriving on site and carrying out work. The clauses governing the situation are 2.7 and 2.6 respectively and the correct term for these directly engaged contractors is 'employer's licensees'. Occasionally, one still comes across references to this clause by its former names of 'artists and tradesmen' or, rarely now, 'the Epstein clause' (this last after the famous sculptor). The contracts term the work done by such contractors 'work not forming part of the contract'. The only sensible interpretation to give to work not forming part of the contract is that it refers to work which the contractor has no obligation to carry out nor for which any right to receive payment.

The clauses are similar in intention although not quite identical in wording. Two situations are envisaged:

- where the contract documents provide for work not forming part of the contract to be carried out by the employer or by persons directly engaged;
- where the contract documents do not make any provision for work not forming part of the contract.

In the first situation the detail in the contract documents must be sufficient to enable the contractor to inform itself precisely of what is to be done and when it is to be carried out. Items of attendance must be specified so that the contractor has the opportunity, at tender stage, to assess the cost to it. Provided this is done, the contractor has no option, as the work proceeds, but to allow the directly employed contractors to do their work.

In the second situation there is no mention of the work in the contract documents, and if the employer wishes to employ others, the contractor's consent must first be sought. The contract stipulates that the contractor must not unreasonably withhold its consent. What is reasonable will depend upon the circumstances. The contractor will require to be informed about the extent of the work, its timing and probably the identity of the person or firm to carry it out.

Reasonable grounds for refusal of consent could be that the contractor has suffered bad experiences working alongside such person or firm in the past or that to allow the work would seriously interfere with progress. The wise contractor will carefully consider all aspects and obtain written assurances from the employer (not the architect) before giving consent.

Clearly, any work done by others during the time the main contract is in progress will cause some disturbance, but there are remedies available which will be discussed later. If work is fully detailed in the contract documents, but the employer wishes to change the work in some way – perhaps to extend its scope – it immediately falls into the second category. There is no provision for varying this work. The variation clauses in the contract refer only to contract work, i.e. that to be executed by the contractor. In these circumstances the employer must seek the contractor's consent before varying the work just as though none of the work was mentioned in the documents.

The wording of the definitions in each contract make clear that directly employed contractors are to be considered not as sub-contractors but as persons for whom the employer is deemed to be responsible (i.e. employer's persons). This means that the employer must make certain that they are properly insured and, if damage occurs through their negligence, the employer will be liable, although it is possible that such liability may be passed on to the directly engaged contractors.

The employer and the contractor must take care that all these employer's licensees have a separate contract with the employer and they must be paid direct by the employer and not, under any circumstances, through the contractor. If the contractor allows itself to become the channel for payment and instructions to such persons, it runs the risk that they may be considered sub-contractors for whom the contractor is responsible.

Disadvantages

In practice, there are few advantages and very many disadvantages for the employer in bringing persons other than the contractor on to the site. Sometimes the employer may wish to do so because, for example, they are the employer's own employees in the case of a local authority or the employer may have a special relationship with them in the case of sculptors, graphic designers or landscape contractors, or it may simply be that the employer wants to have complete control over them, which would not be the case if sub-contractors were named.

Once on the site, or even before entering the site, the employer's licensees will inevitably cause delay and disruption. In the case of a small item such as a specially designed memorial plaque to be fixed complete with unveiling apparatus in the entrance hall of a public building, the disturbance may be so slight as to be insignificant. The work not forming part of the contract, however, may be quite substantial and scheduled to be carried out quite early in the job. Take, for example, the case of special equipment which may necessitate the early supply of baseplates, conduits, etc., which the contractor may have to build in or which the specialist firm insists on fixing itself. If it is late, or even if it is on time, severe disruption may result.

Fortunately, the contract provides remedies for the contractor. This of course is the disadvantage as far as the employer is concerned. The extension of time clause in each contract includes a ground referring to any impediment, prevention or default, by act or omission of the employer (SBC clause 2.29.6, DB clause 2.26.5, IC and ICD clause 2.20.6), and any act of the employer in bringing another contractor on to the site may fall into that category if the directly employed contractor causes any delay to the main contractor.

The situation has changed slightly with the removal of the relevant event which formerly covered this situation. If the work is carefully detailed in the contract documents and the employer's licensees carry it out without any fault, it seems that no ground for extension exists. Of course, all the circumstances must be taken into account. In other cases it seems that granting some extension cannot be avoided. For example, where the work is not previously detailed and carried out only with the contractor's consent, the contractor will certainly be entitled to an extension to cover the carrying out of the work if thereby the completion date is exceeded. Moreover, the impediment, prevention or default clause in the provisions for loss and/or expense (SBC clause 4.24.5, IC/ICD clause 4.18.5 and DB clause 4.20.5) allows the use of directly employed contractors as ground for a financial claim. Since the presence of other contractors on the site will almost invariably cause disruption of regular progress to some extent, the inclusion of employer's licensees, unless detailed in the contract documents and carrying out their work flawlessly, may amount to much the same as a blank cheque from the employer.

The contractor's remedies, and employer's problems, do not end there. If the execution or failure to execute such work causes a delay for a period to be named in the Contract Particulars (clause 8.9.2 in each contract), the contractor may terminate its employment under the contract, with all that entails for the employer (see section 7.2).

Another aspect, which is closely related to employer licensees, is the situation if the employer undertakes to supply any materials or goods for the Works. There is no specific provision in any of the contracts for the employer to undertake direct supply, therefore an additional clause should be inserted if required.

The contractor's letter giving consent under clause 2.7.2 (SBC, IC and ICD) or clause 2.6.2 (DB) is most important and it is suggested that it should be worded somewhat along the following lines:

To the employer

Dear Sir,

We are in receipt of your letter referring to the engagement of [insert name] in accordance with clause 2.7.2 [substitute 2.6.2 when using DB] of the conditions of contract. The work is not detailed in the contracts bills/specification [substitute Employer's Requirements when using DB], but we understand that it will consist of [details] and it will commence on site on [date], reaching completion on [date].

   We give our consent to the work subject to your written confirmation that you recognise our entitlement to extension of time and loss and/or expense in consequence.

Yours faithfully,

It is clear that an employer making use of the provision to employ other contractors on the site is in a very vulnerable position. The employer would be well advised to allow all the work to be carried out through the contractor or, at least, arrange for any additional work to be done after the contractor has reached practical completion stage and left the site. Most contractors would also be well pleased with that arrangement despite generous remedies available in the contract.

MW *and* MWD

MW and MWD make no mention of employer's licensees. Any such arrangement must be with the consent of the contractor and, of course, in this case, it need not be reasonable in withholding its consent. If the contractor does consent, extension of time would be granted under clause 2.7 (2.8 under MWD), but there is no contractual provision for loss and/or expense. It may be difficult to sue at common law because of the difficulty of proving breach of contract, since the contractor, by its consent, has agreed. The answer would seem to be for the contractor to consent on condition that the employer makes special provision to reimburse any loss and/or expense. Powers of termination, however, are available under clause 6.8.2.

3.4 Statutory provisions

Every person or firm has a duty to comply with requirements laid down by statute, that is, by Act of Parliament. It seems clear that a contractor's duty to comply with statutory requirements will prevail over any express contractual obligation.[14]

   Clauses dealing with statutory obligations are now scattered throughout the contract. SBC deals with them in clauses 1.1, 2.1, 2.17, 2.18, 2.21 and 3.7.3, IC/ICD in clauses 1.1, 2.3, 2.15 and 2.16, MW in clauses 2.1, 2.5 and 2.6, MWD in clauses 2.1, 2.6 and 2.7, and DB in clauses 1.1, 2.1, 2.15, 2.16

and 2.18. SBC, IC, ICD and DB provisions are very similar and they will be considered first.

The meaning of 'statutory requirements' is helpfully defined in each contract as being 'requirements of any Act of Parliament, any instrument, rule or order made under any statute or directive, or any regulation or by-law of any local authority or any statutory undertaker' relative to the Works. Statutory instruments are usually made by a Secretary of State, and the most important, from the contractor's point of view, will probably be the Building Regulations, the Scheme for Construction Contracts (England and Wales) Regulations 1998 (Northern Ireland and Scotland have their own versions, which are very similar) and the Construction (Design and Management) Regulations 1994, usually referred to as the CDM Regulations, which are considered below.

The contractor must comply with all statutory requirements and give all notices which may be required by them. It is also responsible for paying all fees and charges legally demandable in respect of the Works. The contractor is entitled to have such amounts added to the contract sum unless they are already provided for in the contract.

Indemnity

SBC and DB, but not IC or ICD, contain indemnity provision, which deserves attention. Not only must the contractor pay charges as above, but it agrees to indemnify the employer against liability in respect of such charges. Therefore, if the contractor fails to pay as legally required, it assumes liability on behalf of the employer. Such liability might well extend to undoing work already done, delays or fines. This is an onerous provision, and one that might easily be overlooked. Its purpose is to keep the employer safe from damage or loss, and its effect would be broader than the position under IC or ICD, where the employer's remedy for the contractor's failure to pay would be to sue for damages for breach of contract.

Indemnity clauses tend to be interpreted by the courts against the person relying on them. In particular, they must be drafted in very precise terms if they are to include indemnity against the consequences of the employer's own negligence.[15] This may give the contractor some relief, but it should note that the time during which it remains liable under an indemnity clause does not begin to run until the liability of the employer has been established, usually by the court.

Divergence

The contractor has a duty to notify the architect immediately if the contractor finds a divergence between statutory requirements and the contract documents or architect's instructions. The contractor has no obligation to search for such divergences (see section 1.1). When the architect receives the notice, or if the architect finds a divergence, SBC clause 2.17.2 and IC/ICD clause 2.15.2 give the architect just seven days to issue an instruction about the divergence. If the architect fails to meet this deadline, the contractor would appear to have clear grounds for an extension of time and reimbursement of loss and/or expense.

Provided that the contractor notifies the architect if it finds a divergence and otherwise carries out the work in accordance with the contract documents and any other drawings and instructions issued by the architect, both forms of contract give the contractor a valuable safeguard. The contractor is not to be held liable to the employer if the Works do not comply with statutory requirements. The contractor is still liable for compliance as far as, for example, the local authority is concerned, but it should be able to recover its costs for rectifying such work from the employer. Thus, if the contractor fails to find a divergence, it is able to escape liability by virtue of this clause.

Emergency

In practice, the contractors should beware of carrying out instructions from anyone other than the architect. If the building control officer directs that work does not comply with the Building Regulations, the only safe course is to refer the matter immediately to the architect. The only exception to this is in the case of an emergency. If the matter really is urgent, the contractor may carry out the necessary work immediately provided that it:

- notifies the architect forthwith of the steps it is taking; and
- supplies only sufficient materials and carries out just enough work to ensure compliance.

The contractor is then entitled to payment for what it has done, just as though the architect had issued an instruction.

MW and MWD

MW and MWD contain broadly similar provisions. The contractor must pay all charges and comply with all statutes, instruments, etc. It must notify divergences which it finds, after which it has no liability for non-compliance as far as the employer is concerned. There is no provision for any indemnity similar to SBC, nor is there provision for emergency compliance. The contractor must, therefore, report any urgent matters to the architect. In circumstances where the matter really cannot wait – for example, if there is an element of danger involved – it is thought that the contractor's obligation to comply with statutory requirements must prevail and it would be entitled to claim reimbursement from the employer for emergency work. In practice, the contractor would have difficulty in persuading the architect that it was not in breach for failing to notify an obvious divergence, or negligence in construction. MW and MWD provisions are quite brief, but adequate for their sort of project.

DB

The provisions of DB broadly follow from SBC, but there are important differences. If either the contractor or the employer finds a divergence, the finder must inform the other. The divergence is between the statutory requirements and the Employer's Requirements or the Contractor's Proposals or any change under clause 5. It is left to the contractor to write to the employer with a proposed amendment to remove the divergence. If the employer consents, the contractor must incorporate the amendment at its own cost. The contract is silent about the situation if the employer does not consent. The employer may not unreasonably withhold or delay such consent. Any dispute as to whether the employer is acting reasonably could be referred to adjudication. This clause has three sub-clauses:

- If a change in statutory requirements occurs after the base date and, in consequence, there must be an addendum to the Contractor's Proposals, it is treated as a change under clause 5. For example, there may be a change in the Building Regulations which requires a partial redesign.
- If a decision is made after the base date by a development control authority (such as a local authority) and the Contractor's Proposals have to be amended, it is treated as a change under clause 5 unless the Employer's Requirements expressly state otherwise.
- Clause 2.15.2.3 states that if the Employer's Requirements specifically state that parts of them comply with statutory requirements, and if such parts later need amending to comply with statutory requirements, the employer must issue a change instruction.

CDM Regulations

Breach of the Construction (Design and Management) Regulations 1994 is a criminal offence; however, except for two instances, a breach will not give rise to civil liability. Therefore, one person cannot normally sue another for breach of the Regulations. Compliance with the Regulations is made a contractual duty so that breach of the Regulations is also a breach of contract. The important clauses are mainly clauses 3.25 and 3.26 in SBC, clauses 3.18 and 3.19 in IC/ICD and clause 3.9 in MW and MWD. The clauses in all the contracts are very similar and it is necessary only to look at SBC 98 in any detail.

Article 5 in SBC assumes that the architect will be the planning supervisor under the Regulations. Under DB, the planning supervisor is assumed to be the contractor. If that is to be the case, careful planning is required to deal with the hand-over after tender stage from the previous planning supervisor to the contractor acting in that role. If one looks carefully, it is possible to see the word 'or' enabling the user to insert an alternative name. It is not at all certain that every architect will want to take on the role. Article 6 records the name of the principal contractor. That will almost always be the contractor under the contract.

There are grounds for termination (failure to comply with the Regulations) in the list in both employer and contractor termination clauses (clauses 8.4 and 8.9).

Three points are very significant and they should be read together. Clause 3.25 of SBC has been included to provide that the employer 'shall ensure' that the planning supervisor carries out all the duties under the Regulations and that where the principal contractor is not the contractor, it also will carry out its duties in accordance with the Regulations. There are also provisions that the contractor, if it is the principal contractor, will comply with the Regulations. The contractor must also ensure that any sub-contractor provides necessary information. Compliance or non-compliance by the employer with clause 3.25 is no longer expressly a relevant event and a relevant matter under clauses 2.29 and 4.24 as it was under JCT 98 clauses 25 and 26. Nevertheless, it may be possible to bring the situation under the impediment, prevention or default relevant event and relevant matter, depending on the circumstances. With careful consideration this may well be a fruitful source of claims for contractors. Every architect's instruction potentially carries a health and safety implication which should be addressed under the Regulations. The Regulations impose a formidable list of duties on the planning supervisor. Most of them are to be found in Regulations 14 and 15. Some of these duties must be carried out before work is commenced on site. If necessary actions delay the issue of an architect's instruction or, once issued, delay its execution, the contractor will be able to claim. The same is true of DB except that the instruction would be from the employer. The contractor's role as planning supervisor is irrelevant for this purpose.

Where, as in DB, the contractor carries out all or some of the design, it takes on the designer's duty under the Regulations, and any question of delay or disruption will require most careful thought.

There may be occasions when the Regulations do not fully apply to the Works as described in the contract. If the situation changes owing to the issue of an architect's instruction or some other cause, the employer may be faced with substantial delay as appointments of planning supervisor and principal contractor are made and appropriate duties are carried out under the full Regulations.

It is certain that the key factor for employers, architects, planning supervisors and principal contractors is to structure their administrative procedures very carefully if they are to avoid becoming in breach of their contractual obligations.

Statutory undertakers

A statutory undertaker is any body, such as an electricity supplier or gas supplier, which derives its authority from statute. There tends to be some misunderstanding of the position of such a body in relation to the contract. SBC clause 3.7.3 and DB clause 3.3.3 make clear that a local authority or statutory undertaker carrying out work as part of its statutory obligations is not a sub-contractor. This appears to be generally understood, and, although not expressly mentioned in IC, ICD, MW or MWD, the position must be the same under those forms also. The result is that the statutory undertaker has no contractual liability to either employer or contractor when it is using statutory powers. If such a body enters the site to lay cables, make connections, etc. within its statutory powers, it may cause severe disruption or delay. All four forms will provide some remedy for the contractor in terms of extension of time, but the contractor cannot contractually recover any loss and/or expense in respect of the disruption either from the statutory undertaker or from the employer.

The position is quite different if the statutory authority is carrying out work on site which is not a statutory obligation. For example, the electricity supplier may be wiring a house as would any electrical sub-contractor. In those circumstances, the authority:

- may have a contract with the contractor and be in the position of a named or domestic sub-contractor under SBC clauses 3.7, 3.8 and 3.9, DB schedule 2 paragraph 2 and clauses 3.3 and 3.4, IC/ICD schedule 2 and clauses 3.5, 3.6 and 3.7, and MW/MWD clause 3.3;
- may have a contract with the employer and be licensee under SBC, IC or ICD clause 2.7 or DB clause 2.6.

In the first instance the contractor's remedy for disruption by the statutory authority would be against the authority itself under the terms of the sub-contract. In addition, depending upon whether the authority was a nominated or named sub-contractor, there would be other important safeguards under the terms of the main contract (see section 3.2). In the latter instance, if the authority were a licensee, the contractor's remedies would be against the employer under the main contract.

It is, therefore, of the utmost importance to the contractor that it correctly identifies the true role of the statutory undertaker. It is generally fairly clear if the undertaker is a sub-contractor, but it may be the subject of dispute whether it is an employer's licensee.[16]

A contractor faced with major losses due to the activities of a statutory authority carrying out its statutory obligations may not be totally without remedy if it can show that its loss was caused by the authority's negligence. In those circumstances the contractor should take legal advice to assess the chances of a successful action in tort.

3.5 Third party rights and warranties

These are very complicated provisions. SBC and DB clause 7A make provision for third parties to obtain rights under the contract. Prior to the introduction of the Contracts (Rights of Third Parties) Act 1999, there was what was known as 'privity of contract'. Only parties to a contract could have any rights under its provisions. Therefore, a contract between two people agreeing to give money to a third could not be enforced by that third person.[17] The Act changed that to give third parties who are mentioned in a contract by name or type or class the right to claim under the contract. Most contracts include a clause to remove the effect of the Act (SBC, IC, ICD and DB clause 1.6, MW and MWD clause 1.5).

However, SBC and DB now take advantage of the Act, and an exception is inserted into clause 1.6 accordingly. The idea is that, to save the need for the contractor to enter into a host of warranties in favour of the funder and each purchaser or tenant, the funder, purchasers and tenants are given rights against the contractor under the contract. In order to achieve this, the rights in question are set out in schedule 5 in SBC and DB. Inevitably, the rights are set out in much the same way as they would occur in a warranty. In order for the rights to be available to any particular third parties, the Contract Particulars must be completed to indicate the name, class or description of the party. For example, the entry may say 'all tenants' in respect of a particular part of the building. There is a great deal more to complete in the Contract Particulars in order to enable this system to work. The procedure is daunting in its complexity, and great care is needed to complete the relevant items properly.

In order to activate the rights, the architect must issue a notice to the contractor. The funder, tenant or purchaser gains the rights on the date of receipt of the notice by the contractor.

If it is decided not to take advantage of the third party rights procedure, there is provision under SBC and DB clauses 7C and 7D to require the contractor to enter into warranties within fourteen days in the usual way provided that details of the relevant funder, tenants and purchasers have been entered in the Contract Particulars. The relevant warranties are CWa/P&T for a purchaser or tenant and CWa/F for a funder. IC and ICD have similar contractor warranty arrangements under clauses 7.5 and 7.6. SBC and DB under clauses 7E and 7F, and IC and ICD under clauses 7.7 and 7.8, also make provision for warranties from sub-contractors. Details of the sub-contractors who will be required to supply warranties are to be inserted in the Contract Particulars. Then these sub-contractors will be expected to provide warranties to the tenants, purchasers or funders listed in respect of the contractor's third party rights or warranties. The relevant sub-contractor warranties are SCWa/P&T for a purchaser or tenant and SCWa/F for a funder. There is provision for the sub-contractor to propose amendments to the warranty subject to approval by the employer and the contractor, who must not unreasonably withhold or delay such approval.

The Contract Particulars may be completed to permit the giving of collateral warranties to the employer by the sub-contractors.

MW and MWD make no provision for warranties of any kind.

References

1 Helstan Securities Ltd v Hertfordshire County Council (1978) 20 BLR 70.
2 St Martin's Property Corporation Ltd and St Martin's Property Investments Ltd v Sir Robert McAlpine & Sons Ltd, and Linden Gardens Trust Ltd v Linesta Sludge Disposals Ltd (1993) 63 BLR 1; Darlington Borough Council v Wiltshire Northern Ltd (1994) 69 BLR 1.
3 Murphy v Brentwood District Council (1990) 50 BLR 1.
4 Dawber Williamson Roofing Ltd v Humberside County Council (1979) 14 BLR 70.
5 ACA Form of Building Agreement (1998 revision).
6 Piggott Construction Co. Ltd v W. J. Crowe Ltd (1961) 27 DLR (2d) 258.
7 Martin Grant & Co. Ltd v Sir Lindsay Parkinson & Co. Ltd (1984) 3 Con LR 12.
8 Piggott Foundations Ltd v Shepherd Construction Ltd (1993) 67 BLR 48.
9 Kitson Sheet Metal Ltd and Another v Matthew Hall Mechanical & Electrical Engineering Ltd (1989) 17 Con LR 116.
10 Henderson v Merritt Syndicates (1994) 69 BLR 26.
11 Holt and Another v Payne Skillington and Another *The Times* 22 December 1995.
12 Haviland and Others v Long and Another, Dunn Trust Ltd [1952] 1 All ER 463.
13 A. Davies & Co. (Shopfitters) Ltd v William Old Ltd (1969) 67 LGR 395.
14 Street v Sibbabridge Ltd (1980) unreported.
15 Walters v Whessoe Ltd and Shell Refining Co. Ltd (1960) 6 BLR 24.
16 Henry Boot Construction Ltd v Central Lancashire New Town Development Corporation (1980) 15 BLR 1.
17 Tweddle v Atkinson (1861) 1 B&S 393.

4 Work in progress

4.1 Setting-out

In an effort to start the job and, having started it, to maintain reasonable progress, the contractor sometimes assumes more than its fair share of responsibility. An example of this may be observed in the contractor's attitude to setting-out when faced with wholly or partially inadequate drawings. Very often, it does its best on the information available when it should be seeking to protect its own position.

Setting-out is covered by SBC clause 2.10, IC and ICD clause 2.9. MW and MWD make no mention of setting-out, but it is thought that a term is to be implied on similar lines to the express terms in the other contracts.

SBC and ICD clauses are virtually identical. The architect's obligation to determine levels and provide setting-out information is, in each contract, made subject to the contractor's obligation to provide the architect with relevant levels and setting-out information for the CDP work. The provision in IC is similar, but obviously excludes any reference to CDP work. It is good to see that the somewhat misleading reference to setting-out at ground level which was contained in JCT 98 has been removed from SBC.

DB has no similar term dealing with setting-out. This is because it is the contractor's obligation, as designer, to fix all the relevant setting-out dimensions and levels. It must be implied that the contractor has the same setting-out duty as if one of the other traditional contracts were used. The employer's obligation is merely to define the boundaries of the site (clause 2.9).

It is the contractor's responsibility, under any of the six contracts, to set out the Works accurately, and if it makes any mistake, it must amend it at its own cost. For example, a contractor may set out a building in such a way that the dimensions of the building work properly in relation to all the other building dimensions, but the building as a whole is in the wrong position on the site.

This kind of error may become obvious only when site works are in progress. Clearly, the contractor may have carried out many thousands of pounds' worth of work by that time. The contractor's obligation seems to be to tear down what it has built and start again, setting it out correctly this time. Certainly, where someone is in breach of contract, the other party is entitled to receive an amount of money which would put it in the same position (so far as money could do it) as if the contract had been correctly carried out.[1] In the case of incorrect setting-out, that would mean the cost of demolition and re-erection. However, this very strict and draconian approach would be modified in practice by the courts or an arbitrator, and if the cost of rectification was out of all proportion to the benefit to be gained, rectification would not be supported and a nominal sum might be awarded instead.[2] All the circumstances would have to be taken into account, including whether the injured party intended to have the matter put right or simply to pocket the money.

If the building encroaches on to neighbouring land, it amounts to trespass and the contractor will be liable for damages if sued directly by the neighbour or as third party if the neighbour sues the employer.

Fortunately, most errors are detected quite early, often while the building is still at the foundation stage, and the damage can be limited. What of the situation where the contractor sets out wrongly so that the employer has substantially more building than expected? There was a case some years ago in which the building, a school, was half a metre longer than shown on the drawings. The error was not discovered until the architect designed a floor tiling pattern and found that it was too short. On the face of things, the employer had gained rather than lost by the contractor's error. In such an instance it would probably be unreasonable to expect the contractor to correct the error, although that is its strict obligation. The contractor would certainly have no claim for reimbursement for additional work and materials used. Moreover, the employer would probably be able to claim that it would be involved in additional maintenance expense for the life of the building.

It is probably to overcome such problems that SBC, IC and ICD provide that the architect, with the employer's consent, can instruct the contractor not to amend errors in setting-out and make an appropriate deduction from the contract sum. There is no guidance in the contract regarding what might constitute an appropriate deduction. The situation is similar to that created when the employer opts not to have defects corrected at the end of the rectification period (see section 7.1), and an appropriate deduction might be the cost to the contractor of rectification of the defective setting-out.[3] It is more likely that a much smaller sum would be indicated. In the case mentioned above, the cost would be substantial and it is doubtful if a court would enforce the point. More likely, the extra cost of maintenance and decoration would form a basis. Each case has to be decided on its own facts and if the contractor's setting-out error really was impossible to live with, demolition and rebuilding could be ordered.

In all cases where errors are accepted by the architect and employer, the amount to be deducted is likely to be a source of argument. It should be noted, however, that once the architect has issued an instruction accepting errors in setting-out on the basis that a deduction is to be made later, the contractor is freed from the threat of expensive corrective measures.

Since the contract is remarkably vague regarding the exact method of calculating what is to be deducted, an astute contractor should be able to keep such deductions within reasonable limits. If the contractor is unhappy about the deduction and the architect refuses to adjust it, the remedy is to refer the matter to the appropriate dispute resolution procedures. Such problems should be capable of quick resolution by adjudication.

A contractor who makes an error in setting-out is often understandably annoyed at the prospect of amending mistakes. It may unjustly accuse the architect or clerk of works of being aware of the error long before the contractor noticed it. However, under the general law the architect has no duty to the contractor to detect errors.[4]

Inadequate drawings

What if the architect's drawings are inadequate? The architect must determine the levels and must provide accurate drawings.[5] The most common failure in this area is that the drawings do not indicate beyond doubt how the building is to be set out. Triangulation from several fixed points is the only certain way of avoiding misunderstandings. It is good practice for the architect to provide special setting-out drawings, shorn of all surplus information and indicating only the important setting-out dimensions.

If the drawings do not contain all relevant information, the contractor should immediately write to the architect requesting the missing details. If the architect does not supply them on time, the contractor has a claim for extension of time and loss and/or expense. The architect cannot escape from obligations by asking the contractor to set out as the contractor sees fit pending inspection by the architect. Although the contractor's course of action is, in fact, inaction until it receives proper drawings, most contractors would try to make progress by setting out a number of pegs for the architect to inspect. If the contractor decides to do this, then, as soon as the architect has approved the setting-out, the contractor should produce a drawing showing the actual triangulated dimensions as set out on site and send it to the architect, requesting confirmation that the setting-out is correct. If the contractor does not protect its interests in this way, it is laying the groundwork for disputes later. A suitable letter could be somewhat as follows:

To the architect

Dear Sir,

We refer to our letter of the [date] in which we informed you that the information on your drawings was insufficient to enable us to set out the Works accurately.

You responded by telephone, asking us to set out the Works to the best of our ability based on the information provided.

We have carried out your instructions and you visited the site on [date] and approved our setting-out.

We enclose our drawing No. [insert] showing the principal dimensions of our setting-out and we should be pleased to receive your written confirmation of approval before we proceed.

If we do not receive such confirmation by return of post, we shall be obliged to notify you of delay to the Works and disruption for which we will seek appropriate extension of time and financial recompense.

Yours faithfully,

The contractor will be involved in a considerable amount of additional work if it elects to attempt to set out from inadequate drawings, and this procedure cannot be recommended. It would not be so bad if the contractor could claim payment for its additional work, but an instruction from the architect to the contractor to, in effect, design the siting of the building is not an instruction empowered by the contract (see section 4.2) except in relation to CDP work.

It cannot be stressed too strongly that many contractors are the authors of their own misfortune. This is not because they are trying to be devious, but quite the reverse: they are trying to maintain progress. In so doing, they often shoulder more responsibility than the contract would impose. A large part of the problem is that the contractor cannot afford delays. Despite what some architects may think, contractors never gain as a result of delay, even if they escape liquidated damages and recover loss and/or expense.

The contractor's ideal is to complete the Works in less than the contract period. Thus, if it simply sits and waits for the full setting-out information, it is losing money. The contractor may be successful in claiming some or all of its losses at some time in the future, but that does not help the contractor to fund the work at the time funding is most needed. It is therefore likely that contractors will continue to try to overcome deficiencies in drawings, particularly setting-out drawings, for no reward. An awareness of the architect's obligations, however, can make the contractor's lot easier.

Of course, under DB and the CDP content of SBC, ICD and MWD, it is the contractor's task to produce drawings which are adequate for its own setting-out, unless of course the employer has misguidedly included all the production drawings as part of the Employer's Requirements.

4.2 Release of information and architect's instructions

The use of an information release schedule applies only to SBC, IC and ICD. Whether or not one is to be provided should be stated in the recital. If the schedule is provided, it should be noted that it is not a contract document. The provision is contained in clauses 2.11 of SBC and 2.10 of IC/ICD. The information – presumably, drawings details, schedules and so on – is to be released by the architect in accordance with the schedule. There is provision for the parties to agree that the information can be released at some other time than shown on the schedule. There is also a saving clause, if the architect is prevented by the action or default of the contractor.

Anecdotal evidence suggests that the schedule is rarely used. The reason is obvious. Any contractor will need the schedule showing the sequence and timing of the information release in order to work out a tender. That means that, before a contractor is appointed, the architect has to decide the order in which the contractor will require the drawings. It is clear that the schedule effectively sets in stone the contractor's programme before it is even produced. Few architects are so brave, or foolhardy, as to do that, quite apart from the financial disadvantage to an employer of a tender based on such artificially fixed criteria.

If the schedule is not provided, or if it does not contain all the information which will be necessary, the architect must provide further drawings and details reasonably necessary to explain and amplify the contract drawings: SBC clause 2.12, IC and ICD clause 2.11. The architect must issue this information and any instructions to enable the contractor to carry out and complete the Works in accordance with the contract. One of the important provisions of the contract is the requirement that the contractor must complete the Works by the date for completion. Therefore, the architect's obligation is to provide further drawings, information and instructions at such a time as to enable the contractor to complete by the completion date.

The clause, however, proceeds to amplify the situation. One might say that JCT is here guilty of gilding the lily, because had the clause been left at this point the meaning would have been relatively clear. However, it goes on to say that the architect must have regard to the progress of the Works or, if the architect thinks that practical completion is likely to be achieved before the contract completion date, the architect should have regard to the completion date. Otherwise, account must be taken when the architect believes it was reasonably necessary for the contractor to receive further drawings, details, etc. This appears to open the door to the architect being able to delay issuing further information and simply to match the contractor's progress, even if the contractor is in delay. The danger here is that a 'chicken and egg' situation will arise so that it may be difficult, with hindsight, to judge whether the architect's late issue of the information is responsible for the contractor's slow progress or vice versa.

If the contractor knows *or* has reasonable grounds to believe that the architect is not aware when the contractor should have a drawing, the contractor must advise the architect sufficiently before the time when the contractor would need the information. It is to do this to the extent that it is reasonably practicable. A master's degree in crystal-ball gazing is probably required for this exercise and no doubt contractors will continue to set out a list of when and what information is required at the beginning of the contract and update it as the work proceeds. The extension of time and loss and expense clauses have been modified and simplified accordingly.

MW has a very simple clause dealing with this topic which simply obliges the architect to issue further information which is 'necessary' to properly carry out the Works.

Obviously, where CDP is used, the contractor is obliged to provide design documents, calculations and any other information necessary to amplify the Contractor's Proposals. This is dealt with by SBC clause 2.9.2, ICD clause 2.10.2 and MWD clause 2.1.5. Under DB, the contractor must provide similar information (clause 2.8). It is important to understand that the provision of this information is not to allow the architect or employer's agent to change it. There is no power to do that. Objection may be made to the information, but only if it does not comply with the contract. Under SBC and DB, schedule 1 sets out a design submission procedure based on marking the submitted drawings in the well-known A, B or C categories.

To some contractors, the position with regard to architect's instructions is simplicity itself. If the architect issues an instruction, the contractor must carry it out and eventually the contractor will be paid for its trouble. Although this is generally true, it is by no means the whole story and some-times it can be quite false. Provision is made for architect's instructions in SBC clauses 3.10–3.14, IC/ICD clauses 3.8–3.11 and MW/MWD clauses 3.4 and 3.5. A very similar provision in clauses 3.4–3.9 of DB provides for employer's instructions.

Leaving MW and MWD aside for a moment, the contractor must forth-with comply with any instruction given by the architect provided that the instruction is empowered by the conditions, that is to say, the other clauses of the contract. 'Forthwith' in this context does not mean 'immediately'; it means 'without delay or loss of time'.[6]

In order to discover what instructions are empowered, it is necessary to search through the whole contract to find the appropriate clauses. For example, there are twenty-one such clauses in IC and ICD. It is important to know what instructions are empowered, because if the contractor simply carries out an instruction which the architect, or the employer under DB, is not empowered to give, the contractor will be in breach of contract and the employer will have no liability to pay. In such cases, the contractor may have recourse against the architect personally or, in the case of DB, it may be able to argue that the instruction, although outside the contract, must still be paid for.

The architect is in the position of an agent with limited authority by virtue of the architect's contract with the employer. The contractor need not worry what the precise terms of that authority are in any particular case. If an instruction is empowered by the contract, the contractor may – indeed, must – carry it out even though the architect may not, in fact, have the employer's consent.

A good example is the situation where an architect issues an instruction for the carrying out of additional work. The architect's agreement with the client will probably stipulate that the architect must obtain the client's consent before issuing an instruction of that nature. But if the architect issues such an instruction, it is empowered by the contract, and the contractor is entitled to be paid for it. If the architect has not obtained the client's consent, that is a matter between architect and client and no concern of the contractor, which is entitled to rely only on the contract between contractor and employer.

In SBC, IC and ICD there are two exceptions to the requirement that the contractor must comply forthwith. The first concerns the issue of instructions requiring a variation of an obligation or restriction imposed by the employer in the bills or specification on working space, access to site, working hours or the execution of the work in a specific order. In such cases the contractor need not comply, provided that it makes reasonable objection in writing to the architect. A reasonable objection might well be, for example, that the contractor would be quite unable to make satisfactory progress under the new conditions. If the architect agrees, the instruction may be withdrawn, but any dispute may be referred to immediate adjudication, arbitration or litigation as appropriate.

The second exception arises if the contractor is unsure whether an instruction is empowered by the contract. It may request the architect to specify the clause. The architect must specify the clause in writing forthwith. On receipt of the architect's notification, the contractor may either immediately invoke the dispute resolution procedures on the point or accept the clause nominated by the architect and carry out the instruction. If the contractor elects to carry out the instruction, it is 'deemed' to have been empowered even if it is not in fact.[7] This is a useful protection for the contractor and ensures that, after such enquiry and response, the contractor is entitled to the benefits of other contract provisions in respect of the instruction – for example, the valuation of variations, extensions of time and loss and/or expense – even if the architect has made a mistake.

SBC contains another exception if the instruction requires the contractor to provide a schedule 2 quotation. In such a case the variation must not be carried out unless the architect has issued a confirmed acceptance or another instruction without requiring a quotation.

Under DB or where the CDP is used, a further exception applies, and an instruction which adversely affects the contractor's design may be the subject of the contractor's written objection.

Whether the contractor is entitled to refuse to comply with any instructions in the exceptions mentioned pending the result of adjudication or arbitration is a difficult question. Whether it is a good idea will depend upon circumstances and, if in doubt, advice should always be sought. The contractor is certainly entitled to refuse compliance if its original objection is well founded. This calls for remarkable powers of foresight, because if the result of adjudication or arbitration were that the original objection was unjustified, the contractor would be in breach of contract if it had not complied. If it were thought prudent to carry out the instruction before the result of arbitration of the matter is available, it would be wise to make it clear that the instruction was being carried out without prejudice to the reference. At that stage, an appropriate letter would be drawn up by the contractor's contractual adviser who is dealing with the dispute resolution procedure.

Compliance notice

If the architect, or the employer under DB, considers that the contractor is unreasonably delaying carrying out a valid instruction, he has the power to issue a notice requiring compliance within seven days of receipt. If after seven days the contractor has not complied, the employer (note, not the architect) may employ others to do the work and an appropriate deduction may be made from the contract sum. It is difficult to see how this can be any other than the additional cost to the employer of getting another contractor to do the work. It may also include additional architect's and quantity surveyor's fees, etc. If the architect's instruction is valid, there is very little the contractor can do except carry out the instruction as soon as it receives such a notice.

If the costs involved are likely to be substantial and the contractor considers the architect's insistence on compliance within seven days unreasonable, it can always write a strongly worded letter which may convince the architect or serve as useful evidence of prevailing circumstances and intentions in any future adjudication or arbitration on the point. In practice, architects are usually, and properly, reluctant to bring others on to the site to do the work unless relations have deteriorated considerably, the work has little prospect of being completed or the project is almost completed.

Oral instructions

SBC clause 3.12.1 states that all instructions must be in writing and then, illogically, states what is to happen if instructions are given orally. Basically, oral instructions are of no effect. Contractors often do, but should never, comply with them. If such an instruction is given, the contractor must confirm it in writing to the architect within seven days. The architect then has a further seven days in which to decide whether to dissent. If the architect does not dissent, the instruction takes effect from the expiry of the latter seven days. From the date of the oral instruction, anything between eight and about fifteen days can elapse before the instruction becomes effective.

There is really no excuse for oral instructions, and if the contractor makes it clear – in a friendly way, of course – that it will not act on oral instructions, the architect may stop giving them. SBC and DB go on to provide, in clause 3.12.3 and 3.7.3 respectively, that the architect, or the employer under DB, may confirm instructions within seven days, and in such a case the contractor need not itself confirm, which seems to be stating the obvious – although how the contractor is to know that the architect will confirm on, say, day 6, when its own obligation is to confirm, is not stated.

If neither architect nor contractor confirms within the time period stated, but the contractor carries out the instruction, the contractor has no rights in the matter. The architect, or the employer under DB, may, however, confirm the instruction at any time up to the issue of the final certificate and it will be deemed to have taken effect on the date it was issued orally. The moral is clear: all instructions must be in writing. If not, the contractor must immediately confirm, wait seven days for dissent and, if there is none, carry it out.

Although an architect may argue that there is no alternative but to give an oral instruction if it is required immediately and the architect is at some distance from site, such a contention carries little weight now that facsimile machines are the norm. An instruction sent by fax is properly served in writing.[8]

IC and ICD make no provision for oral instructions. They are not mentioned. Therefore, confirmation by the contractor will have no effect. Confirmation by the architect will have the same effect as if the instruction had been issued at the time of confirmation.

However, an architect who relies upon the absence of a written confirmation in order to avoid authorising payment later may have a nasty shock in store. Looked at in terms of what the contract says, the position is clear. Work done which is not properly instructed under the terms of the contract is work 'not in accordance' with the contract, and the architect has a duty to order such work to be removed from site. The architect who does not so order is in breach of his contractual duty to the employer. Moreover, the contractor is in breach of its contract if it does not remove it. It appears that a contractor who carries out work on the basis of an architect's oral instruction, which the architect later refuses to confirm, is entitled to amend the work to restore it to its previous condition in order to avoid a breach of contract. There is good authority, however, to the effect that compliance with an architect's instruction issued orally is a good defence to claim for damages for such breach of contract.[9]

On the other hand, a contractor who accepts an oral instruction under IC or ICD and carries it out will not be able to argue for any reason that it was not empowered by the contract.[10]

MW and MWD

The position under MW and MWD is somewhat different from that under the other four forms. Instructions must be in writing. If oral instructions are issued, they must be confirmed by the architect within two days. There is no provision for confirmation by the contractor. The contractor must carry out architect's instructions forthwith, and there is provision for the architect to issue a seven-day compliance notice and for the employer to engage others to do the work if the contractor defaults. The actual clause may appear to be all-embracing in its effect, because it states that the architect 'may issue written instructions'. There is no mention of the instructions being empowered by the conditions (as in SBC, DB, IC or ICD).

It may seem that the architect can issue instructions about any matter that the architect sees fit. However, the architect must act within the scope of his or her authority. The courts usually take a dim view of provisions in contracts which appear to give one party unlimited powers and they construe them narrowly, having in mind the main object of the contract and limiting the provisions accordingly.[11]

This clause only gives the architect power to issue instructions regarding the Works included in the contract. Five other clauses relate to specific instructions:

- not to make good defects at his own cost (2.10 (2.11 under MWD));
- to order variations (3.6);
- to expend provisional sums (3.7);
- to exclude persons from the Works (3.8);
- to reinstate and make good after loss by fire, etc. (5.4B.2).

In addition, the architect probably has power to instruct postponement, correction of inconsistencies, opening up and testing, and removal of defective work.

Issue of instructions

The issue of instructions, unless omitting work, will provide grounds for extension of time under all four contracts. In addition, under SBC, DB, IC and ICD, the issue or late issue, if requested by the contractor in reasonable time, will be grounds for a financial claim under clauses 4.23 (SBC), 4.19 (DB) and 4.17 (IC/ICD). Under the provisions of MW and MWD it is stipulated that when the architect values an instruction under clause 3.6, the valuation must include any loss and/or expense incurred by the contractor in complying with the instruction, under that form alone, and the contractor need make no special application. It is the architect's duty to include the loss and/or expense.

There are one or two popular misconceptions about instructions under traditional contracts. Only the architect has power to issue an instruction, and, as noted above, the power is limited. Neither the clerk of works nor any other consultant may issue instructions, although in practice they often do. An employer who attempts to issue an instruction is in breach of contract. Alternatively, the employer may be considered to be attempting to negotiate a new contract to carry out the work in the instruction.

If the contractor carries out an instruction issued by anyone other than the architect, it does so at its peril. It is not entitled to payment and it is probably in breach of contract itself by so doing. For example, if a contractor carries out an instruction given by the heating consultant to install an extra radiator, the architect can require the extra radiator to be removed from site as not being in accordance with the contract, and the contractor must bear the whole cost (see the discussion above on the confirmation of oral instructions). Obviously, under DB, it is the employer or the employer's agent who may issue instructions.

What is an instruction?

The golden rule for contractors is clear. They should not carry out instructions unless they are:

- given by the architect, or the employer under DB, in writing or otherwise confirmed;
- empowered by the contract;
- identifiable as instructions.

This raises the question, what is an instruction? The RIBA and ACA have published standard instruction forms which are in common use. The forms are boldly headed 'Architect's Instruction'. It is good practice to issue all instructions on such forms; however, it is not the form but the substance which determines whether it is an instruction, and an instruction can be issued as a letter.

Sometimes the contractor may simply receive a drawing and a compliment slip. It is dangerous to treat this as an instruction unless there is some message on the compliment slip saying that the work shown on the drawing is to be carried out. Anything sent with a compliment slip should be treated with suspicion. If the contractor receives a copy of the employer's letter asking for something to be done, it is not an instruction, except under DB, but an invitation to the contractor to do that something at its own cost.

An item in the minutes of a site meeting probably becomes an instruction when the minutes are agreed as accurate at a subsequent meeting, and possibly before that if the architect, or the employer under DB, is responsible for producing the minutes. One architect regularly scribbled instructions on site on pieces of plywood and the backs of roofing tiles. Such conduct is rather precious, but the instructions are valid, provided they are signed and dated. The production of copies may be a problem.

Not all instructions imply payment. Many instructions may be simply clarifications or may be issued under some provision which has no financial implication, such as SBC clause 3.18.1 requiring the removal from site of work not in accordance with the contract.

An instruction requiring a variation is not valid, and there is no liability on the employer to pay if it is issued in respect of something which the contractor must do anyway in carrying out its obligations under the contract.[12] In passing, it may be noted that an instruction signed in the architect's name by another person is quite valid, provided that the person had the architect's authority to sign.[13]

4.3 Clerk of works

A good clerk of works can make a tremendous difference to work on site in terms of defects, efficiency and general atmosphere.

SBC, IC and ICD make provision for the appointment of a clerk of works. DB, MW and MWD make no such provision, possibly in the case of DB because the employer's agent may take this role. In the case of MW and MWD, it is probably because projects executed under this contract are expected to be relatively small and uncomplicated. There is no reason, however, why provision for a clerk of works should not be made in the specification for either contract.

SBC clause 3.4 states that the employer may appoint a clerk of works who is to be under the direction of the architect. The only duty of the clerk of works is to inspect, and the contractor must give every reasonable facility. In practice, that means that the contractor should not hinder the clerk of works, who must be allowed access to all parts of the Works. The contractor cannot be expected to go to great lengths to erect scaffolding to enable the clerk of works to inspect the building, but if the scaffolding is already in position, the clerk of works must be able to use it.

Directions

If the clerk of works gives a direction to the contractor, it is of no effect. This means that not only can the contractor ignore it but also that it is in breach of contract if it takes any action, or becomes inactive, as the case may be, on account of the direction. The clause then contains the curious provision that a direction of the clerk of works will be effective if given on a matter on which the architect is empowered by the contract to issue instructions and if the direction is confirmed by the architect within two working days of issue of the direction. The direction is then deemed to be an architect's instruction effective from the date of the architect's confirmation.

Two points arise from this provision. First, the architect's confirmation must be given within two working days – an action which rarely happens. If confirmation is delayed, the direction is not deemed to be an architect's instruction. But does it matter, since the confirmation itself presumably ranks as an instruction anyway? Second, even if confirmed within two days, the direction is effective only from the date of confirmation. There seems little point in the clerk of works giving a direction in the first place. It would seem more appropriate if the clerk of works simply gave the architect a telephone call and requested the issue of an instruction about the matter.

The usual scenario is that the clerk of works issues a direction. The contractor need not, in fact must not, take any action on it no matter how urgent. The contractor may feel certain that it will be confirmed and either does the work and takes a chance or possibly holds back from that portion of the work until it hears from the architect. If the contractor takes the latter course, it may suffer disruption and/or delay for which it has no redress.

If it takes the former course, the architect may not confirm and the contractor will not be paid for any additional or substituted work. Worse, the architect may issue instructions, in accordance with clause 3.18.1, that the work done in response to the direction of the clerk of works is not in accordance with the contract and must be removed from site. The moral is quite clear: if given a direction by the clerk of works, the contractor should ignore it, carry on working as usual and, if confirmation is received from the architect, then, and only then, carry out the direction. If, as a result, the work is disrupted, the contractor will be able to make application for loss and/or expense.

Clearly, this is not the way a contractor usually works. Directions from the clerk of works are quite often treated as though they are architect's instructions from the moment of issue.

IC/ICD clause 3.3 is much shorter. It merely makes reference to the clerk of works being appointed by the employer as an inspector under the direction of the architect. The clerk of works is not empowered to give directions, even directions which may be ignored! From this point of view IC and ICD appear eminently sensible. Some clerks of works have expressed the view that they should be empowered to issue instructions under the contract. Such provision would obviously lead to confusion on site.

The duty of the clerk of works is to the employer. The clerk of works has no duty towards the contractor to find defects. The approval of the clerk of works counts for nothing. It may or may not be indicative of the architect's attitude. If the clerk of works points out defects, the contractor would be wise to take notice, but there is no substitute for a competent site agent or, as the contract says, person-in-charge.

The contractor will often present daywork sheets to the clerk of works for signing. This is incorrect practice. SBC clause 5.4 and DB clause 5.7 stipulate that such vouchers must be delivered to the architect, the architect's authorised representative or the employer respectively, for verification before the end of the week following the week in which the work has been carried out. The clerk of works is not the architect's or the employer's representative unless the architect or the employer specifically states the same in writing to the contractor.

A negligent clerk of works will not remove liability from the architect so far as the employer is concerned, but if the employer successfully sues the architect for negligence, the damages payable by the architect may be reduced on account of the negligence of the clerk of works.[14] This is a point of more interest to the architect and the employer than to the contractor. It depends on the clerk of works being employed by the employer and not by the architect, as sometimes happens, and raises the point of the employer's vicarious liability for the actions of the clerk of works.

Defacement

Many clerks of works are in the habit of using chalk or wax crayon to deface work or materials which they consider to be defective. In this they are sometimes encouraged by the architect and, indeed, the contractor itself. Some contractors, however, quite rightly take exception to such conduct and protest to the architect. If materials are defaced by the clerk of works it is because, presumably, they are not in accordance with the contract. Either they will be removed by the contractor or the architect will instruct removal. They are, therefore, the contractor's property. It may be that the contractor could use materials in other less stringent situations except for the fact that they are defaced. There is no doubt that the clerk of works is not entitled to deface Works or materials. The clerk of works' role is purely to inspect. The contractor should put a stop to defacement as soon as it first appears. This is best done by a suitably worded letter on the following lines to the architect:

Dear Sir,

It is common practice for the clerk of works to deface work or materials considered to be defective. The basis for such action presumably is to bring the defect to the notice of the contractor and ensure that it cannot remain without attention.

We object to the practice on the following grounds:

1   The work or materials so marked may not be defective and we will be involved in extra work and the employer in extra costs in such circumstances.
2   The work or materials so marked, if indeed defective, will not be paid for and will be our property when removed. We may be able to incorporate it in other projects where a different standard is required. Defacement by the clerk of works would prevent such re-use.

We will take no point about the defacing marks we noted on site today, but if the practice continues, we will seek financial reimbursement.

Yours faithfully,

Specialist clerks of works

It is the practice of some large organisations, which maintain a body of clerks of works on their permanent staff, to allow a number of so-called 'specialist' clerks of works to inspect the site at regular intervals. It is not always appreciated that the contractor has possession of the site under licence from the employer and, with the exception of the persons noted in the contract, it has the right to refuse admittance to the site. The situation is clearly delicate, but the contractor would be ill-advised to allow a multitude of inspectors to roam the site. Of course, it is always open to the architect, or the employer under DB, to make such 'specialist' clerks of works authorised representatives, in which case the contractor must allow them access, certainly under SBC, DB, IC or ICD provisions (clause 3.1).

The architect is unlikely to make large numbers of persons authorised representatives because it would imply that they can act for the architect, issue instructions, etc. The situation is not absolutely clear, but any contractor faced with 'specialist' clerks of works might try refusing access until such time as the position has been clarified in writing.

Snagging

A 'snag' sounds much less concerning than a 'defect'. The term 'snag' is not used in any contract, but common usage suggests that it is just another name for a defect. Although it may suggest a minor defect, no such distinction is made in practice.

A clerk of works will often issue snagging lists, particularly as practical completion draws near. If the contractor accepts such lists as merely helpful reminders of work to be done, all should be well. Problems and disputes sometimes arise on site because the clerk of works produces snagging lists, and after items receive attention the architect produces another list. Contractors often feel aggrieved about it, maintaining that one list is quite enough. In fact, the snagging list is not of contractual significance unless issued by the architect and it is clearly stated that it represents the only points requiring attention before the architect is prepared to issue a practical completion certificate. It is the contractor's obligation to complete the building in accordance with the contract, and the clerk of works can never absolve the contractor from that obligation. The clerk of works has no duty to produce snagging lists.

A clerk of works can be very useful but it should always be remembered that the clerk of works is just an inspector empowered to look and to note, and that is all. If the clerk of works issues directions, the contractor should not act on them, unless of course they relate to obvious defects, until they are confirmed. Confirmation by the contractor is not effective.

References

1 Robinson v Harman (1848) 154 ER 363.
2 Forsyth v Ruxley Electronics & Construction Ltd (1995) 73 BLR 1.
3 William Tomkinson & Sons Ltd v Parochial Church Council of St Michael (1990) 6 Const LJ 319.
4 Oldschool and Another v Gleeson (Contractors) Ltd and Others (1976) 4 BLR 103.
5 London Borough of Merton v Stanley Hugh Leach Ltd (1985) 32 BLR 51.
6 Hudson v Hill (1874) 43 LJCP 273.
7 For an interesting judicial comment about 'deeming' see: Re Coslett (Contractors) Ltd, Clark, Administrator of Coslett (Contractors) Ltd (In Administration) v Mid Glamorgan County Council [1997] 4 All ER 115.
8 Hastie and Jenkerson v McMahon (1990) *The Times* 3 April 1990.
9 G. Bilton & Sons v Mason (1957) unreported.
10 Bowmer & Kirkland Ltd v Wilson Bowden Properties Ltd (1996) 80 BLR 131.
11 Glyn and Others v Margetson & Co. and Others (1893) AC 351.
12 Sharpe v San Paulo Brazilian Railway Co. (1873) 8 Ch App 597.
13 Anglian Water Authority v RDL Contracting Ltd (1992) 27 Con LR 76.
14 Kensington and Chelsea and Westminster Area Health Authority v Wettern Composites Ltd (1984) 1 Con LR 114.

5 Money

5.1 Payment

SBC provides for the issue of ten different kinds of certificate. The figure for IC, ICD, MW and MWD is six (no certificates are issued under DB, because there is no independent architect as certifier). Many of these certificates are not financial, although most contractors associate the word 'certificate' with money. In order to be a valid certificate, a document must be headed 'Certificate', start with the words 'I certify' or be clearly referenced to the contract provision empowering issue. In any event, it must be the clear expression of the judgment, skill or opinion of the architect.[1]

Financial certificates are covered by SBC clauses 4.6–4.15, IC and ICD clauses 4.3–4.14, MW and MWD clauses 4.3–4.8. SBC, IC and ICD provisions are similar and will be discussed together. MW and MWD will be considered separately below. What follows can be no more than a brief summary and the parties are advised to study the contract terms carefully.

Interim certificates are to be issued at the dates noted in the Contract Particulars, usually at monthly intervals. They must state the amount due to the contractor from the employer, and the employer has fourteen days from the date of the certificate in which to pay. Failure to honour certificates within the due time is a ground for suspension or termination by the contractor (see section 7.2) and entitles the contractor to interest. The architect may ask the quantity surveyor to carry out a valuation not more than seven days before the date of the certificate, but responsibility for the correctness of the sum on the certificate remains with the architect.[2] Alternatively, the contractor may submit an application for payment not later than seven days before certification date, which the quantity surveyor may either accept or reject by sending the contractor the quantity surveyor's own view in the same detail as submitted by the contractor. The amounts to be included in each certificate are as follows. Subject to retention:

- total value of work properly executed;
- total value of materials delivered to site for incorporation, if not premature;
- value of listed off-site materials;

Not subject to retention:

- in respect of payments or costs due in regard to opening up and testing, loss and/or expense, statutory obligations, provisional sum insurance, insurance premiums;
- in respect of restoration, repair or replacement following certain insured loss or damage;
- fluctuation payments;

less reimbursement of advance payment and any authorised deductions.

The certificate should express the amount due to the contractor as the balance between the total of the above amounts and the amounts already certified on previous certificates.

What if the architect simply refuses to issue a certificate? If the contractor can show that the refusal is due to interference by the employer, the contractor is entitled to terminate its employment. Under JCT standard form contracts the contractor cannot recover money without a certificate. That is because where there is an arbitration clause the courts take the view that the contractor's remedy for the absence of a certificate is to seek arbitration.[3] In the rare instance of a building contract not containing an arbitration clause, it is likely that the contractor could sue without a certificate.[4] In general, the architect's failure to issue a certificate at the proper time is a breach of contract for which the employer is liable.[5]

If the contractor can show that the architect deliberately interfered with the contract in order to keep the contractor out of money, the architect may be personally liable to the contractor for the tort of interference with contractual rights.[6] It is not something which is easily proved.

It is often thought that the architect named in the contract must sign all certificates for them to be valid. That is not correct. It is quite sufficient if a properly authorised person signs in the name of the named architect.[7] It matters not that the person is not an architect, provided the person is authorised by the architect to sign in the architect's name. It should be noted, however, that the architect's signature on a certificate does not amount to issuing the certificate. To accomplish that under the provisions of these contracts the architect must at least bring the contents of the certificates to the attention of the employer.[8]

Payments under DB

Valuations may be made in accordance with clause 4.13 (alternative A) or 4.14 (alternative B). Alternative A refers to stage payments. The contractor is entitled to payment of the amounts stated in the Contract Particulars when the stages of work set out are completed. It is important that the stages are described very carefully so that there is little room for dispute. In addition to the stage payments, the contractor is also entitled to such other payments specified under the contract as become due – for example, payment of the value of change instructions under clause 5 and of loss and/or expense under clause 4.19.

Alternative B provides for periodic payments at intervals stated in the Contract Particulars. If no intervals are stated, the period is one month. Essentially, this type of valuation is the usual one, consisting of the value of work properly executed, the value of change instructions, loss and/or expense and so on.

The procedure for application for payment is very simple and it is contained in clause 4.9. In the case of alternative A, applications are to be made on the completion of each stage. It seems that applications for the value of changes carried out and loss and/or expense must wait until the next stage application, even if completion of the next stage is a considerable time distant. The last stage assumes practical completion and the next stage is on the issue of the notice of making good under clause 2.36.

Applications under alternative B are to be made at the intervals stated in the Contract Particulars up to the date after practical completion occurs. From then on, further applications can be made whenever further amounts are due and on the issue of the notice of making good defects.

Applications under this form of contract are contractor-driven. By that I mean that the contractor states the amount it has calculated as being due and the employer must either pay or ensure that an effective proposed payment notice or withholding notice is served.

Clause 4.9.3 allows the employer to state in the Employer's Requirements precisely what details the contractor must include with the applications. A wise employer will make sure that the contractor is required to provide a proper breakdown with each application. In practice, it is surprising how seldom that requirement is included.

Payments under MW and MWD

MW and MWD provisions stipulate that the architect must certify progress payments to the contractor at four-weekly intervals. It is possible that the parties may come to some other arrangements in the case of a small job, such as stage payments, although this is no longer formally acknowledged in the contracts. The employer must pay within fourteen days of the date of issue of the certificate. Each certificate must state the value of Works properly executed and materials on site less amounts previously certified. The retention, which is not expressed as being held in trust, is normally 5 per cent. If the employer becomes insolvent, the contractor's retained money is not protected and the contractor is not entitled to demand that the money be kept in a separate bank account. Half the retention is to be released within fourteen days of the certificate of practical completion.

Priced activity schedule

Priced activity schedules apply to SBC (clause 4.16.1.1), IC and ICD (clause 4.7.1.1) if the second, fourth or fifth recital respectively so states. The activity schedule must be attached to the contract with each activity priced so that the total of the prices equals the contract sum, less provisional sums, prime cost sums, contractor's profits on this and the value of any work for which approximate quantities are included in the bills of quantities. The contractor has an option as to whether it wishes to provide an activity schedule. If it does, the schedule is used to assist in the valuation of interim certificates. The prices in the activity schedule are to be proportioned so as to give the amount in the interim certificate. This may mean that the contractor will receive sums in the interim certificates which more closely reflect the work carried out.

A priced activity schedule is usually used as an alternative to priced bills of quantities. It seems strange, at first sight, to use it as well as bills of quantities. It should be noted, however, that even where a priced activity schedule is used, bills of quantities are still used for the valuation of variations.

Advance payment

SBC, DB, IC and ICD make provision for advance payment, but all contain a note in the Contract Particulars excluding local authorities. The employer can pay a sum to the contractor on a date stated in the Contract Particulars. The idea is that the sum will be paid early, in order to assist the contractor to finance the project. The contractor must reimburse the employer the amounts, and at the intervals, which the two parties agree and state in the Contract Particulars. In these circumstances the contractor is required to give a bond, usually in the form reproduced at the back of the contract (SBC clause 4.8, DB clause 4.6, IC and ICD clause 4.5).

Retention

A percentage, usually 5 per cent, is held by the employer, who may use the money whenever the contract directs that the employer may deduct from money due or to become due to the contractor. This retention fund is also useful at the end of the job to ensure that defects are made good. Except under MW and MWD, the retention money is held on trust for the contractor. That means that, although it is held by the employer, it really belongs to the contractor, which is why the employer has only limited powers of access to it. It is settled that the contractor has the right to demand that the retention money be paid into a separate bank account, clearly named in favour of the contractor.[9]

Indeed, SBC and DB include a provision to that effect, although not if the employer is a local authority. The importance of keeping trust money separate is that if the employer becomes insolvent, clearly identified trust money can be paid to the contractor. Otherwise, the contractor has to take its chances along with other creditors. The employer has no right to use the money for his or her own purposes, and to do so would be a breach of trust. The employer has an obligation to keep the trust money in a separate bank account even if there is no clause to that effect or if such a clause has been struck out,[10] and that, of course, also applies to local authorities. The contractor is not obliged to make a request for the money to be placed in a separate account each time a certificate is issued and paid. It may make just one request at any time.[11] If the employer goes into receivership before a separate account has been set up, it seems that the contractor has lost its right to the money.[12]

The contract states that the employer has no obligation to invest the money. In other words, the contractor is not entitled to interest. Since there is a statutory obligation to invest, which a contract provision cannot defeat, it is possible that a contractor who insisted would be entitled to interest. Note, however, that no contractor has, so far, felt confident enough to test the point through the courts.

There is no retention clause, as such, in IC and ICD, the amount of retention being dealt with under clause 4.9 in a positive, rather than a negative, way by stating the percentage to be paid rather than the percentage to be withheld. The amount to be paid under this clause includes release of half the percentage withheld. The remaining retention is withheld until the issue of the final certificate. The whole of the retention, clause 4.10, is stated to apply only where the employer is not a local authority. Therefore, if the employer is a local authority, the position is not clear and the clause should be amended.

The second part of the clause allows the employer to use the fund for money which the contract allows the employer to withhold or deduct. If this clause does not apply to local authorities, perhaps the intention is that the retention will not be a trust fund and, therefore, the local authority does not need a special provision to be able to use the fund. How that operates in practice only time will tell. SBC provides for release of the second half of retention after the certificate of making good is issued.

Under SBC clause 4.19 it is possible for a bond to be provided in lieu of retention if the appendix so states. If the contractor fails to provide a bond, the ordinary retention provisions apply until the bond is provided. There are complicated provisions to take care of the situation in which the bonded sum falls below the figure which would have been in the retention fund.

The final account

Under SBC, IC and ICD, not later than six months after practical completion the contractor must send to the architect, or to the quantity surveyor if so directed, all documents necessary for working out the final contract sum. Not later than three months thereafter, the quantity surveyor must prepare a statement of all valuations of variations and send the contractor a copy of the computations of the adjusted contract sum. SBC provides detailed guidance on the way in which the contract sum is to be adjusted.

The rules for the issue of the final certificate vary between SBC and IC/ICD. In the case of SBC the final certificate must be issued not later than two months from the latest of the following:

- end of the rectification period;
- issue of certificate of making good;
- date the statement and ascertainment is sent to the contractor.

It is common experience for the last event to be the determining factor.

In the case of IC and ICD, issue must take place within twenty-eight days of sending to the contractor the computations of the adjusted contract sum or the issue of the certificate of making good, whichever is the later.

In the case of SBC, IC and ICD the final certificate must state the adjusted contract sum, the amounts previously certified and the balance expressed as either a sum due to the employer or a sum due to the contractor. The employer or contractor, as the case may be, must pay within twenty-eight days of the date of issue.

Unless adjudication, arbitration or other proceedings have been commenced by either party within twenty-eight days of the date of issue, the final certificate is conclusive evidence that:

- where quality or standards of materials, goods or workmanship are expressly stated to be a matter for the architect's satisfaction, the architect is satisfied;
- the contract terms requiring adjustment of the contract sum have been correctly applied;
- all due extensions of time have been given;
- reimbursement of loss and/or expense is in final settlement of all matters under clause 4.2.3 (IC and ICD clause 4.17).

The only exceptions arise in the case of fraud or accidental inclusions or exclusions of work, materials or figures in any computation or arithmetical error. It is clearly stated that no other certificate is conclusive evidence that work or materials are in accordance with the contract. The current position so far as the first item is concerned has been discussed in section 1.2.

Under MW and MWD, the contractor must supply within the specified period, usually three months, from practical completion all documentation for computation of the final certificate. The final certificate must be issued within twenty-eight days of receipt of the documentation, provided that a certificate of making good has been issued. The final certificate is not stated to be conclusive as regards any matter, not even the amount.

All six contracts are lump sum contracts. The contractor is entitled to payment provided it completes substantially the whole of the work.[13] The fact that interim payments are made does not alter the position. Failure to perform substantially means that the contractor cannot recover.[14]

Reference in the contracts to work 'properly executed' refers to work which is in accordance with the contract. The contractor is not, of course, entitled to be paid for defective work. If the architect does certify defective work, the architect may become personally liable if the contractor becomes insolvent.[15]

The arrangements for final payment under DB are somewhat different. Within three months of practical completion the contractor must submit the final account and final statement to the employer for agreement. It becomes conclusive about the balance due to the contractor within one month of the latest of:

- the end of the rectification period;
- the date in the notice of completion making good;
- submission of the account to the employer;

except to the extent that the employer disputes it (clause 4.12.4). If the contractor does not submit the account within three months, the employer, after giving two months' notice, can do it himself. It is then for the contractor to dispute it, if it wishes, within one month of the events noted above.

Either party has twenty-eight days from the date when the account would otherwise become conclusive in which to start adjudication, arbitration or other proceedings. Otherwise, the final account and final statement are conclusive about the same three matters noted earlier in respect of SBC, IC and ICD, i.e. workmanship, extensions of time, and loss and/or expense, when the account becomes conclusive about the balance due. The catch is that if the employer or the contractor, as appropriate, decides to dispute any part of the account, the contract contains no mechanism to make the account conclusive in the future.

Off-site materials

It should be particularly noted that if the employer wishes to pay for materials off site, the materials will have been listed and attached to the contract bills of quantities. The items on the list must be divided into two categories: the first category is 'uniquely identified items' and the second category is 'not uniquely identified items'. Uniquely identified items are those materials or goods which are easy to recognise, such as heating boilers or sanitary fittings or the like. Items not uniquely identified cover such items as bricks, sand, tiles, timber and anything which it would be difficult to recognise as belonging to a particular site.

The architect is obliged to include all such listed materials and goods in the valuation in any certificate and they must be included before delivery to site. Under DB, they are to be included in the contractor's application for payment. This is subject to the contractor fulfilling certain specified criteria. They are as follows:

- The contractor must have provided reasonable proof that it owns the items, so that after it has been paid the value of items included in certificates, the ownership of items will pass to the employer.
- The contractor must have provided the employer with a bond if so required by an entry in the Contract Particulars. The surety for the bond must be approved by the employer and, unless otherwise agreed, the terms of the bond must be as agreed between the JCT and the British Bankers Association. In the case of not uniquely identified materials, the provision of a bond is mandatory.
- The items must be in accordance with the contract.
- They must be kept at the premises where they have been manufactured or stored and either they must be set on one side or, alternatively, they must be clearly marked so as to identify the employer and that they are destined for the Works.
- The contractor must have provided reasonable proof that the items are insured in respect of specified perils.

There is no provision for payment for off-site materials under MW and MWD.

Set-off

The question of set-off has long been a bone of contention. In general, it appears that there is a right of set-off unless, looking at the contract as a whole, it is excluded expressly or by necessary implication.[16] The point is important because, at one time, the architect's certificate was considered as good as cash and had to be honoured.[17] It now seems clear, however, that an employer with good grounds can withhold payment on a certificate and, resisting summary judgment, go to arbitration or trial.[18] This is bad news for contractors, particularly because it appears that the employer need only show that there are reasonable grounds to challenge the certificate.[19] Following the coming into force of the Arbitration Act 1996, it will be rare for contractors to obtain summary judgment if employers fail to pay, and contractors may be wiser to use the contractual power to terminate their employment.

Following the Housing Grants, Construction and Regeneration Act 1996, all four contracts contain express provisions to deal with set-off or, as the contracts now refer to it, withholding or deduction (SBC clauses 4.13.3, 4.13.4, 4.13.5, 4.15.3, 4.15.4 and 4.15.5; IC/ICD clauses 4.8.2, 4.8.3, 4.8.4, 4.14.2, 4.14.3, 4.14.4; MW/MWD clauses 4.6.1, 4.6.2, 4.6.3, 4.8.2, 4.8.3

and 4.8.4; and DB clauses 4.10.3, 4.10.4, 4.10.5, 4.12.8, 4.12.9 and 4.12.10). Because the provisions were inserted as a result of legislation, they are substantially the same in each contract.

The employer must issue a written notice to the contractor within five days of the date of issue of each certificate (including the final certificate) or, under DB, from the receipt of an application for payment or from the date the final account and final statement become conclusive. The notice must state the amount which the employer proposes to pay, to what it relates and how it is calculated. Presumably, the notice will state the amount in the certificate (DB: application) although it is possible, but not certain, that the employer could use the opportunity to abate the sum (i.e. to reduce it, possibly because work has not been done).

If the employer wishes to withhold or deduct any amount from the sum due, including abatement, the employer must issue a written notice not later than five days before the final date for payment of any certificate (or application under DB).[20] This notice must state the grounds for withholding and the amount to be withheld for each ground. If the employer does not give a written notice in accordance with one and/or both provisions, the sum must be paid in full. Under the Act the first notice will suffice for both if it indicates a deduction and sets out the grounds in sufficient detail. It appears from the wording that the contractual position may be the same, but the wise employer will always give both notices if wishing to deduct any amounts, but possibly in the same envelope before the earlier deadline. Obviously, the employer who intends to pay the full amount may omit to give any notices. Technically it would be a breach but, provided the certificate was paid in full, the contractor could hardly complain.

Formerly, the amount due under DB if no withholding notice was served was the amount in the application. The position is now changed. Clause 4.10.5 makes clear that if no withholding notice is given to the contractor, the amount the employer must pay is the amount stated in the first notice (the employer's statement of the amount proposed to be paid). If the employer omitted to give even the first notice, the amount to be paid would be the amount calculated under clause 4.8. This appears, effectively, to give the employer the right to substitute the employer's own valuation for that of the contractor. However, it should be noted that, in the case of the final payment, the absence of both notices results in the payment of the amount in the contractor's or the employer's (as appropriate) final account and the final statement.

SBC clauses 4.13.6 and 4.15.6, IC/ICD clauses 4.8.6 and 4.14.5, MW/MWD clauses 4.4 and 4.9, and DB clauses 4.10.6 and 4.12.11 provide that the employer must pay simple interest at 5 per cent above Bank of England base rate if the employer fails to pay the amount due by the final date for payment. This is in addition to the contractor's other rights to suspend or terminate or, in appropriate cases, to accept repudiation under the general law.

5.2 Variations

Contractors sometimes ask if they are obliged to carry out an architect's instruction requiring a variation. The answer is 'yes' as far as JCT contracts are concerned, but there are some points to note. The clauses authorising the architect to issue instructions requiring variations are SBC clause 3.14, IC/ICD clause 3.11 and MW/MWD clause 3.6.1. In the absence of these clauses, the architect would have no power to order variations, and any attempt to do so by architect or employer would amount to a breach of contract and the contractor might be justified in negotiating a new price for the contract. This is because SBC, IC, ICD, MW and MWD are lump sum contracts and, basically, the contractor has tendered a price for the whole of the work.

There are some other clauses which authorise variations in specific circumstances (SBC clause 2.14.3, IC/ICD clause 2.13.1, etc.), but the principal clauses are the ones noted above. SBC, IC and ICD provisions are similar and will be considered first. Variations are defined in these contracts, in clause 5.1 in each case, as the alteration or modification of the design or the quality or quantity of the Works from that shown in the contract documents. They include: additions, omissions and substitutions; alterations of kinds or standards of materials; or the removal from site of materials delivered for the Works, unless the removal is because the materials are defective. That is all very clear and gives the architect wide powers.

An important provision is the architect's power to issue instructions imposing, adding to, omitting from or altering obligations or restrictions relating to access, working space, working hours or the order of the work. This seems to be a dangerous extension of the architect's powers into the contractor's own field, and there is only one safeguard: the contractor can make reasonable objection under SBC clause 3.14.2 or IC/ICD clause 3.11.2.

The architect may not omit work, which has been measured in the bills and priced by the contractor, in order to give it to others to carry out.[21] The prohibition also covers the omission of provisional sums to allow others to do the work.[22]

The position under DB is broadly similar. An important difference is that variation instructions under clause 3.9 may be given only by the employer. Moreover, they are referred to as 'Change instructions', because the employer may only instruct a change in the Employer's Requirements which may, in turn, require an alteration in the design, quality or quantity of the Works. Indeed, the contractor has the right to refuse to put a change instruction into effect if it has sufficient reason such as its effect upon the design.

The employer must issue instructions regarding the expenditure of provisional sums, but only if they are in the Employer's Requirements. Provisional sums which are in the Contractor's Proposals, but which have never been transferred to the Employer's Requirements, cannot be expended.

The employer has the same power to impose or vary restrictions as under SBC.

Valuation

Under SBC, IC and ICD, variations are to be valued by the quantity surveyor in accordance with the rules laid down in the contract. This procedure may be set aside if the contractor and the employer (not the architect) agree. This allows the employer, for example, to accept the contractor's quotation for a proposed variation. Alternatively, under SBC, there is provision for a contractor's quotation, requested under clause 5.3.1 and schedule 2 (a schedule 2 quotation). In IC and ICD there is simply provision for an agreement on the valuation. The effect of this is discussed later in this section. The rules for the valuation of variations are sensible. In SBC, reference is to be made to the prices in the priced bills. In IC and ICD, reference is to be made to the prices in the priced document (see section 1.1).

Omissions are to be valued at the rate in the bills. Additions and substitutions which are similar in character and executed under similar conditions and without significant change in quantity to work in the bills are to be priced using those rates. If there are changes in the conditions and quantity, valuations are to use the bill rates as a basis. If the work is not of a similar character, or involves other than additions, omissions or substitutions, or if it is not reasonable to value it using bill rates as a basis, then a fair valuation must be made. Rules are set out for the valuation of work on a daywork basis if it cannot properly be measured.

The above is only a general summary, and all parties are advised to study the relevant clauses carefully. It should be remembered, however, that the quantity surveyor has a fair degree of discretion regarding the method of valuation and is under no obligation to obtain the contractor's agreement. The amount to be included in any certificate in respect of such valuation is for the architect to decide,[23] even though, in practice, the architect will usually adopt the quantity surveyor's valuation. Even in adopting the quantity surveyor's valuation, however, the architect has a duty to the client to carry out sufficient checks to be satisfied that the certified sum is correct. That is not to say that the architect is obliged to revalue the work – that would be needless duplication – but the architect should ask for sufficient supporting information from the quantity surveyor so that the work that has been valued can be easily seen. The rules for valuation, where approximate quantities are involved, and the implications of defined and undefined provisional sums should be carefully studied where *Standard Method of Measurement* seventh edition (SMM 7) is to apply.

Important provisions require that if an instruction causes substantial changes in the conditions under which other work is carried out, such other work will be treated as if varied. For example, if the architect issues an instruction changing 'dry lining' to 'three coats wet plasterwork', the contractor will be entitled to be paid for that variation as appropriate. By virtue of SBC clause 5.9 or IC/ICD clause 5.5, it will also be entitled to additional payment because altered conditions may lead the contractor to carry out some of its other operations in a different order or may necessitate added protection. The change in conditions must be substantial.

The variations clauses do not seem to include valuation of the effect of a variation upon the regular progress of the work. Any claims under this head must be made in accordance with SBC clause 4.23 or IC/ICD clause 4.17. In SBC and IC/ICD clauses 5.10.2 and 5.6.2 respectively, it is clearly stated that no allowance must be made in the valuation for any effect on regular progress or other loss and/or expense for which the contractor would be reimbursed under any other clause. If, however, the contractor can show that it would not be so reimbursed under another provision, it is entitled to have the amount valued under the variation clause.

There is an alternative valuation procedure in SBC. It allows the architect to instruct the contractor that a variation is to be dealt with under schedule 2. The contractor has twenty-one days from receipt of the instruction or requested further information to submit its quotation to the quantity surveyor. The quotation must include the value of the work, any extension of time, loss and/or expense and the cost of preparing the quotation. The architect may request other information such as a method statement. Acceptance must be within seven days from receipt and be confirmed by the architect. If not accepted, valuation is to be carried out under the normal rules or the architect must instruct that the variation is not to be carried out. The great advantage of this procedure is that otherwise contentious claim matters are settled at the time of the variation. Points for contractors to watch:

- The contractor has seven days after receipt of the instruction to give written notice that the information provided is not sufficient.
- If a later variation is instructed to schedule 2 work, neither the normal rules nor schedule 2 apply. Instead, the quantity surveyor must make a fair and reasonable valuation based on the earlier schedule 2 quotation.

DB

Valuation under this contract is dealt with under clause 5. The process is similar to SBC except that no architect or quantity surveyor is involved. Effectively, the valuation is to be undertaken by the contractor. If the employer disagrees, the remedy is to serve the first or the first and second notices when the contractor's application for payment is received or, alternatively, to carry out another valuation in accordance with clause 4.8. The standard valuation rules are set out in clause 5.4.

If, in the Contract Particulars, the supplemental provisions are said to apply, provision S4 governs the valuation of change instructions. Essentially, the process is in substitution for clause 5. It is triggered by the issue of an instruction from the employer. If the employer, or the contractor, believes that a valuation, extension or time or loss and/or expense will be entailed, the contractor has fourteen days to submit an estimate of the value, extension of time, loss and/or expense and details of the resources required and a method statement. The employer has ten days to agree or he must:

- instruct compliance and S4 will not apply; or
- withdraw the instruction.

If the instruction is withdrawn, the contractor is entitled to be paid for any abortive design work. The sting in the tail is that if the contractor fails to submit the required estimates, the valuation is to be carried out under clause 3.9, extensions of time under clauses 2.23–2.26 and loss and/or expense under clause 4.19, but no payment is to be made until the final account and final statement, and no financing charges can be included.

MW and MWD

Under MW and MWD, variations are empowered by clause 3.6. It is a very short clause allowing the architect to vary the Works by addition, omission or change or to change the order or period in which they are to be carried out. The contractor has no right of reasonable objection, but can always refer any dispute to arbitration.

Valuation of variations may be carried out by the architect and contractor reaching agreement before the work is carried out. Indeed, clause 3.6.2 states that the architect and the contractor shall endeavour to do so. The standard procedure is valuation by the architect, on a fair and reasonable basis, using the prices in the priced specification, schedules or schedule of rates where relevant. The architect's view as to whether the particular priced document is 'relevant' in any particular instance will, doubtless, prevail, at least until adjudication. The contractor is not in a strong position and, even if priced schedules are used, the employer does not warrant that they are correct. A major difficulty, from the contractor's point of view, is that it is deemed to have included in its price for carrying out and completing the Works in accordance with the contract documents. Thus, work shown on the drawing, but not in the schedules, does not rank as a variation, as would be the case under SBC with Quantities edition. (See also section 6.2 in relation to the inclusion of loss and/or expense.)

MW and MWD no longer make provision for a quantity surveyor to be appointed. A quantity surveyor may be appointed and the architect will probably delegate the carrying out of valuations.

Points

All four forms make reference to the fact that no variation will vitiate or invalidate the contract. This is superfluous, because nothing empowered by the contract can invalidate it. It must not be thought, however, that the architect or the employer under DB can order variations with impunity. If the variation or the sum of all the variations on a particular contract is such that the whole scope and character of the work are changed, the contractor may be entitled to negotiate a totally new contract. The change must be such that the contractor can say that the project is not substantially that for which it originally tendered. For example, if virtually every detail is changed little by little so that flat roofs are replaced by pitched, baths by showers, small wooden windows by large metal windows and so on, the contractor may well have a case. It should be noted that the mere number of variations is not important and each situation must be considered on its merits.

All variations, except omissions, entitle the contractor to extensions of time if they cause the completion date to be exceeded. They also entitle the contractor to be reimbursed for direct loss and/or expense.

A situation which sometimes arises concerns work which the architect insists is included in the contract, but which the contractor is equally certain should be an extra. Certain ground works sometimes fall into this category. The architect may refuse to authorise a variation. What is the contractor to do? The work may be substantial in quantity and value. If the contractor is confident of its position, its remedy appears to be to refuse to carry out the work unless a variation is authorised and, failing such authorisation, to treat the contract as repudiated and sue for damages.[24]

This may not always be a prudent action to pursue, the contractor risking the possibility of bearing huge losses if the result of legal action is not in its favour. If the contractor proceeds with the work and attempts to claim later, it may be considered that it has carried out the work on the architect's interpretation of the contract and that it is entitled to no extra payment. There is no easy answer but, at the very least, if the contractor elects to proceed, it should make its position clear in writing to the employer and do the work 'without prejudice' to its right to claim later.

In some circumstances it may be held that the employer has implicitly promised to pay if the work is, in fact, additional to the contract.[25] In practice, it should be possible for the parties to agree that the contractor carries out the work on the clear understanding that the matter can be settled by adjudication or arbitration and payment made if the work is found to be a variation. A suitable letter might run along the following lines:

To the Employer

SPECIAL DELIVERY – WITHOUT PREJUDICE

Dear Sir

We refer to your letter of [date] and ours of [date] relating to [describe work]. It is our firm view that this work is not included in the contract and, therefore, constitutes a variation for which we are entitled to payment.

We are advised that we can refuse further performance until you authorise a variation. If you continue to refuse so to authorise, we may treat the contract as repudiated and sue for damages.

Without prejudice to our rights, we are prepared to carry out the work, leaving this matter in abeyance for future determination by adjudication or arbitration, if you will agree in writing and confirm that you will not deny our entitlement to payment in such reference if, on the true construction of the contract, the work is held to be not included.

Yours faithfully,

Dayworks

The valuation of dayworks is often a contentious issue. If the work is to be valued by measurement or on some other basis, the submission of signed daywork sheets by the contractor will be irrelevant. It is only if the work is to be carried out and valued on a daywork basis that signed daywork sheets assume any importance. If the sheets are signed by the architect's authorised representative, they must be valued using the hours and resources on the sheets. The quantity surveyor has no power to go behind the sheets and substitute another estimate of the hours it should have taken to do the work.[26] Signing under the phrase 'for record purposes only', which is very common, does not imply that there is power to value the work on a different basis.[27] Even where sheets have not been signed, they will be good evidence of the hours spent, provided that they were completed by the contractor at the end of the relevant day in accordance with the architect's or quantity surveyor's requirements.[28]

References

1 Token Construction Co. Ltd v Charlton Estates Ltd (1973) 1 BLR 50.
2 R. B. Burden Ltd v Swansea Corporation [1957] 3 All ER 243.
3 Lubenham Fidelities & Investments Co. Ltd v South Pembrokeshire District Council and Wigley Fox Partnership (1986) 6 Con LR 85.
4 Panamena Europea Navigacion Ltda v Frederick Leyland & Co. Ltd (1947) AC 428.
5 Croudace Ltd v London Borough of Lambeth (1986) 6 Con LR 70.
6 Edwin Hill & Partners v First National Finance Corporation plc [1988] 07 EG 75.
7 London County Council v Vitamins Ltd [1955] 2 All ER 229.
8 London Borough of Camden v Thomas McInerney & Sons Ltd (1986) 9 Con LR 99.
9 Rayack Construction Ltd v Lampeter Meat Co. Ltd (1979) 12 BLR 30.
10 Wates Construction (London) Ltd v Franthom Property Ltd (1991) 53 BLR 23.
11 J. F. Finnegan Ltd v Ford Sellar Morris Developments Ltd (1991) 53 BLR 38.
12 Macjordan Construction Ltd v Brookmount Erostin Ltd (1991) CILL 704.

13 Hoenig v Isaacs [1952] 2 All ER 176.
14 Bolton v Mahadeva [1971] 2 All ER 1322.
15 Sutcliffe v Thackrah [1974] 1 All ER 319.
16 Gilbert-Ash (Northern) Ltd v Modern Engineering (Bristol) Ltd (1973) 1 BLR 73.
17 Dawneys Ltd v F. G. Minter Ltd (1971) 1 BLR 16.
18 C. M. Pillings & Co. Ltd v Kent Investments Ltd (1985) 4 Con LR 1.
19 R. M. Douglas Construction Ltd v Bass Leisure Ltd (1991) 53 BLR 119.
20 Whiteways Contractors (Sussex) Ltd v Impresa Casteli Construction UK Ltd (2000) 16 Const LJ 453.
21 Carr v J. A. Berriman Pty Ltd (1953) 27 ALJR 273; Commissioner for Main Roads v Reed & Stuart Pty Ltd (1974) 12 BLR 55; Vonlynn Holdings Ltd v Patrick Flaherty Contracts Ltd (1988) unreported.
22 AMEC Building Ltd v Cadmus Investments Co. Ltd (1996) 13 Const LJ 50.
23 R. B. Burden Ltd v Swansea Corporation [1957] 3 All ER 243.
24 Peter Kiewit Sons' Co. of Canada Ltd v Eakins Construction Ltd (1960) 22 DLR (2d) 465.
25 Molloy v Liebe (1910) 102 LT 616.
26 Clusky (trading as Damian Construction) v Chamberlin (1994) April BLM 6.
27 Inserco Ltd v Honeywell Control Systems Ltd (1996) unreported.
28 JDM Accord Ltd v Secretary of State for the Environment, Food and Rural Affairs (2004) 93 Con LR 133.

6 Claims

6.1 Extension of time

The basic idea of the extension of time and liquidated damages clauses in SBC, IC, ICD, MW, MWD and DB is very simple and straightforward. If the contractor does not complete the Works by the date for completion, the employer is to be paid pre-agreed damages. If, however, the contractor is delayed owing to certain specified events, the contract period will be extended, thus releasing the contractor from the obligation to pay damages for the overrun in respect of those events. In essence, that is all there is to the provision. In practice, however, the application of the clauses seems to cause problems out of all proportion to the issues at stake. Part of the difficulty lies in the number of myths and misunderstandings which cloud the sensible operation of the contract provisions.

The extension of time provisions are for the benefit of the contractor and the employer. It is easy to see that the contractor benefits from an extension of time because it releases it from the obligation to pay liquidated damages. The benefit to the employer is a little more complex.

Under SBC, IC and ICD clause 2.4 and DB clause 2.3 the employer must give possession of the site to the contractor on the date for possession stated in the Contract Particulars. The contractor is obliged to commence the Works on the date for possession and regularly and diligently proceed with them so that they are complete on or before the date for completion stipulated in the contract. MW clause 2.2 and MWD clause 2.3 are to much the same effect, but without the requirement to proceed regularly and diligently (albeit that failure to do so is a ground for termination under clause 6.4.2). Under SBC, IC, ICD and DB, but not under MW or MWD, the employer may defer possession by up to six weeks.

It is important to remember that under the general law the contractor's obligation to complete the Works by the contractual completion date is removed if the employer or the employer's agents are responsible for some or the whole of the delay.[1] Such actions as the issue of instructions or any kind of interference or obstruction fall into this category. In such cases time becomes 'at large'; that is to say that there is no longer any date by which the contractor must complete and, therefore, no date from which liquidated damages can be calculated.[2] The contractor's obligation is then to complete the Works within a reasonable time. A reasonable time may, of course, be the length of the original contract period plus the period for which the employer has caused delay. The employer may still claim damages, but instead of merely being able to deduct them, the employer is faced with the problem of having to prove them in court.

The above general rules may be amended if there is an express term in the contract.[3] There is such an express term in the six contracts under consideration which allows the architect or the employer under DB to grant an extension of time for employer's defaults and thus preserve the employer's right to deduct liquidated damages for any period of overrun beyond the extended date. The benefit to the employer is now clear.

The extension of time provisions are contained in clauses 2.26–2.29 of SBC, 2.19 and 2.20 of IC/ICD and 2.23–2.26 of DB. (The provisions in MW and MWD are somewhat different and contained in clauses 2.7 and 2.8 respectively, which will be considered later.) The provisions clearly take account of two distinct types of delay:

- delays caused by the employer (these are the most important);
- delays caused by events outside the control of either the contractor or the employer.

The 2005 JCT contracts now helpfully list them in that order.

An extension of time may be given only if the event falls within the events contemplated by the extension of time clause. This means that if the employer causes delay by some default which is not included in the clause, an extension of time will not be able to be given and time will become 'at large'. In deciding whether there is power to extend time and whether the event does fall within the events in the clause, the court will construe the terms against the person wishing to rely upon them. The clause must provide expressly or by necessary inference for an extension on account of the employer's fault.[4]

There may be faults of the employer which will not fall within the terms of SBC, DB, IC or ICD clauses and which, therefore, will cause the liquidated damages clause to be inoperative. Even more interesting is the position under MW clause 2.7 and MWD clause 2.8. Until 1988 there was no list of events for which an extension might be granted, but only the bald statement that the architect could make a reasonable extension of time for delays caused by 'reasons beyond the control of the contractor'. A number of authorities considered that this clause was not specific enough to embrace any faults of the employer. If correct (the point has not, to my knowledge, been tested in the courts under the Minor Works contracts), any additional instructions or other delaying actions or omissions by the employer would make time 'at large'. The point appears to have been taken by the Joint Contracts Tribunal, which added the words 'including compliance with any instruction of the Architect under this contract whose issue is not due to the default of the Contractor'.

A further sentence was added to make clear that reasons within the control of the contractor include defaults by any person or firm employed by the contractor, e.g. sub-contractors. This was in response to a judicial decision which appeared to indicate that in certain circumstances the actions of a sub-contractor are beyond the contractor's control.[5] The current MW/MWD contracts retain these features. Architect's instructions are likely to be the main source of delay attributable to the employer. It is not, of course, the sole example of possible employer-generated delay, and the earlier comments are probably still valid in respect of other such delays, for example late information. Indeed, it is thought that the addition of just one example of an employer's delay gives the argument greater force. In the majority of MW/MWD contracts the amount of liquidated damages at risk is probably too small to be worth the expense of testing the point.

A contractor must make a separate application for loss and/or expense. An extension of time has no financial implication other than release from liquidated damages, not even the much beloved 'additional preliminaries'. A contractor may make application for reimbursement of direct loss and/or expense with or without an extension being granted. The courts have emphasised that a claim for loss and/or expense does not depend on a prior grant of extension of time.[6] The contractor may make such financial claim even if it completes the work before the contract date for completion, provided only that it can satisfy the architect that it has suffered loss and/or expense due to circumstances which may or may not also be a ground for extension of time.

It can readily be appreciated that this approach is to the contractor's financial advantage in that it is not bound in any way to the grounds for extension of time when claiming financially.

Contractor's duties

It is for a contractor to prove a link between an event and a period of delay. It is not sufficient for the contractor merely to show a delay and point to an event, in effect saying, 'It must have been that.'[7]

The contractor has very precise duties under SBC clause 2.27 and DB clause 2.24. The contractor must submit a notice in writing as soon as it thinks it is being delayed or that it is likely to be delayed. It must do this whether the delay is the contractor's own fault or due to some other reason. The purpose of this provision is clearly to enable the architect or the employer under DB to take whatever steps may be appropriate at the earliest possible moment to minimise the effect of the delay. If the contractor fails to give such notice, it is in breach of contract.[8] Failure to give notice, however, may not remove the obligation to grant an extension in appropriate circumstances. There has been some discussion as to whether the minutes of a site meeting constitute effective notice. On balance, it is considered that a specific written notice is required[9] (which of course may be in the contractor's report handed to the architect at the meeting).

The notice must include the reasons for the delay and state which of the causes is one of the relevant events listed in clause 2.29 (clause 2.26 under DB). As soon as possible, at the same time as the notice if it can, the contractor must send an estimate of the length of the delay beyond the date for completion. The effect of each relevant event must be noted separately, stating whether the delays will operate concurrently. If the contractor thinks the completion date will not be exceeded, it must say so at this point. The contractor must send further notices to update the original notice as necessary.

The contractor has two more duties. It must constantly use its best endeavours to prevent delay and it must do everything reasonably required by the architect or the employer under DB to proceed with the Works. This obligation has been defined as doing 'everything prudent and reasonable' to achieve an objective.[10] This is not, as sometimes thought, an acceleration clause. Neither does it give any power to order the contractor to incur additional cost to maintain progress in spite of delays. If it did, the extension of time clause would be redundant. It simply means that the contractor must take all reasonable steps to reduce the effect of a delay; for example, by redeploying part of the workforce to other sections of the work. In practice, it may mean little more than continuing to work regularly and diligently as far as practicable. Acceleration of the Works can be achieved only by agreement between contractor and employer. If the architect or the employer under DB makes a reasonable suggestion to progress the Works, the contractor should take notice, but it is not required to incur any additional expense.

It must be stressed that failure by the contractor to carry out its duties meticulously will not remove the obligation to grant an extension in appropriate instances, but the contractor's failure may be taken into account in deciding on the appropriate extension in any particular case. Probably the yardstick is whether and to what extent the architect or the employer could have taken action to reduce the effect upon the completion date if the contractor had given notice at the correct time. The principle is that a party is not entitled to gain as a result of its own wrong.[11]

The provisions in IC/ICD clause 2.19 are in broadly similar terms although not so detailed. The obligation to give notice of any delay, to state its cause and to use best endeavours is repeated, but the contractor is not obliged to make an estimate of the extent by which the completion date will be exceeded. Obviously, a prudent contractor will do so and will supply a very full and well-documented case.

The provisions of MW clause 2.7 and MWD clause 2.8 are very short. The contractor's obligation is to notify the architect in writing if it becomes apparent that the Works will not be completed by the date for completion for reasons beyond the contractor's control. That is the end of its duties in respect of delays, but if it is wise, the contractor will present a full and detailed statement of the causes and an estimate to the architect.

Architect's duties

The architect's duties under MW and MWD are equally briefly stated. After receiving the contractor's notice, the architect must make, in writing, a reasonable extension of time. There is no time limit set, but the architect should act promptly. In general, the decision should be made before the contract completion date is reached, but there may be instances when the delay is continuing. In such a case the architect may wait until the delay has ended before making an extension.[12]

SBC and DB give a time limit of twelve weeks from the receipt of required particulars from the contractor in which an extension of time must be given. If there is less than twelve weeks to the contract completion date, the architect (employer under DB) must endeavour to reach a decision and notify the contractor before the completion date even if the decision is that it is not fair and reasonable to make any extension. In giving an extension, any instruction which requires an omission of work may be taken into account and the contractor must be so notified. Which of the relevant events has been taken into account and the time allocated to each must also be stated. Two points must be considered: whether the delay is caused by a relevant event and, if so, whether it is likely to delay completion.

IC and ICD place no time limit on the exercise of the architect's duty. A fair and reasonable extension is to be made as soon as he is able to estimate the length of delay beyond the completion date. The architect must, however, make the decision within a reasonable time, having regard to all the circumstances. This contract specifically allows the architect to grant an extension of time for certain events in clause 2.20 which occur after the date for completion and which are the fault of the employer. The absence of this provision would lead to such events, occurring after completion, rendering time 'at large'. SBC does not contain this provision and it is by no means certain that the architect would have the same power if it were not for the provision noted in the next paragraph. The kind of event referred to includes such actions as the architect issuing instructions for extra work after the date for completion has passed but while work is still in progress. Such an event would clearly mean further delay.

A valuable provision in SBC and DB allows the architect or the employer under DB to make good any deficiencies in previous extensions. After the date for completion has passed, the architect or the employer may, but no later than twelve weeks after practical completion must, either:

- fix a later completion date than previously fixed; or

- fix an earlier completion date after taking into account omissions from the work; or
- confirm the completion date previously fixed.

A date, however, may not be fixed earlier than the original contract completion date. This provision enables account to be taken of any relevant event which the contractor has failed to notify and which is an employer default, thus preserving the employer's right to deduct liquidated damages.

IC/ICD contain a provision for the architect to make further extensions of time at any time up to twelve weeks after practical completion. There is no provision for reducing any extension previously granted, but otherwise this term serves the same purpose as the term mentioned above in SBC. It is sometimes stated that the twelve-week period is not mandatory and the architect may carry out his review after that time.[13] Architects would be prudent, however, to ensure that they carry out their duty in this respect within the stipulated period and not rely upon a legal decision which depended upon the special circumstances in which the employer was trying to obtain an advantage from the architect's default. A subsequent decision confirms the twelve weeks deadline.[14]

Relevant events

An extension of time may be given only on grounds listed in the contract. SBC, DB, IC and ICD refer to these grounds as 'Relevant Events' (clauses 2.29, 2.26 and 2.20 respectively). The wording in the three contracts is very similar, but there are some important differences. It may be helpful to consider each event separately, indicating where there is a significant difference. In most instances here the wording has been shortened and simplified.

Variations   This includes any other things which the contract states are to be treated as a variation or requiring one.

Compliance with architect's instructions or employer's instructions under DB   This is a most important clause because it relates to something within the control of the employer. An extension must be granted if the employer is to preserve the right to deduct liquidated damages. It should be quite clear whether an event is a compliance with an instruction, but it should be noted that an instruction to open up work for inspection under clause 3.17 (SBC), 3.12 (DB) or 3.14 (IC and ICD) will give grounds for an extension only if the uncovered work is found to be in accordance with the contract.

The contractor's work in accordance with the special clause 3.15 of IC and ICD, in which the contractor must propose action to ensure that there are no similar failures, will also provide grounds for extension of time on the same basis, although the work itself is to be carried out at no cost to the employer. An architect's instructions under SBC clause 3.18 or the employer's instructions under DB clause 3.13 have similar effect. (This provision is discussed in section 1.4.) Instructions regarding the expenditure of provisional sums are also included except where they relate to defined work under SMM 7 which is not relevant to DB (*Standard Method of Measurement* seventh edition) when the contractor is deemed to have made due allowance in programming and planning in accordance with rule 10.4. (Note that if full information is not provided in the bills of quantities in accordance with rule 10.3, a correction must be made under clause 2.14.1 of SBC and the correction will be treated like a variation instruction, i.e. the contractor will be entitled to extension of time and to make application for loss and for expense.)

Deferment of possession   This clause occurs in SBC, DB, IC and ICD. Under MW and MWD, failure by the employer to give possession to the contractor on the date stated in the contract is a serious breach for which the contractor may be able to claim substantial damages.[15] SBC, DB and IC/ICD sensibly allow the employer to defer possession for a limited time (clauses 2.29.3, 2.26.3 and 2.20.3 respectively) and equally sensibly allow the architect to grant an extension.

Approximate quantities   The contractor is entitled to an extension of time where approximate quantities are included in the bills of quantities and the approximate quantities are not a reasonably accurate forecast of the quantity of the work required. This clause is not included in DB.

Suspension by the contractor of performance of his obligations   This clause puts into effect the extension of time to which the contractor is entitled under section 112(4) of the Housing Grants, Construction and Regeneration Act 1996 after it has exercised its right to suspend for non-payment (see section 7.2).

Impediment, prevention or default by the employer   This clause is to ensure that there is no event for which the employer is responsible which causes a delay to the contractor, but for which the architect has no power to issue an extension of time. This is a very broadly worded clause. Advantage is taken of the broad scope of this clause to omit a number of relevant events which were in the 1998 contracts and thus simplify the list. Among the former relevant events now covered by this ground are:

- architect's failure to comply with the information release schedule or to provide information at the right time;
- work not forming part of the contract;
- employer's failure to give access to the site;
- compliance or non-compliance with the CDM Regulations.

Reference to delay on the part of nominated sub-contractors has been removed, and also the inability to secure labour or materials.

The carrying out of work or failure to carry out work in pursuance of its statutory obligations, by a statutory undertaker in relation to the Works This clause relates to the laying of pipes and cables, connections, etc. which the body has an obligation to carry out. The work must relate to the contract Works in order to qualify. Thus, if the electricity supplier causes delay to a site solely because of its operations in regard to a neighbouring site, the contractor has no grounds for extension.

Exceptionally adverse weather condition This is somewhat broader in meaning than the old term 'exceptionally inclement weather conditions' . Adverse weather can be rain, snow or frost. It can also be very hot and dry conditions or even high winds. The key word is 'exceptionally'. Thus, heavy snow in January may not be exceptional. In order for this clause to operate the weather must be exceptional having regard to the time of year or the location of the site. It is usual for the architect to require meteorological reports for perhaps ten years prior to the occurrence.

The contractor is, of course, expected to exercise a reasonable degree of anticipation at tender stage. Therefore, if the date for possession is stated as 25 November, the contractor must allow for the kind of weather which can be expected at that time of year when planning its site operations. In other words, it must expect, for example, snow in December, frost in February, etc. Note that it is the adverse weather which must be exceptional, not the period of time. The contractor's actual progress must be considered, even if the contractor is late through its own fault.[16]

Loss or damage caused by any one or more of the specified perils This is quite straightforward. The risks (the contract refers to them as 'specified perils') are listed in the contract. They include fire, lightning, explosion, storm, tempest, etc. If progress is delayed thereby, there should be no problem in obtaining a suitable extension. It should be noted, however, that the risks are more restricted than the wider 'all risks'.

Civil commotion, use or threat of terrorism or activities of the authorities in consequence A civil commotion is more serious than a riot.[17] Regrettably, terrorism is self-explanatory.

Strike, lock-out, etc. affecting any of the trades employed on the Works or engaged in the preparation, manufacture or transportation of goods for the Works or, where CDP or DB is concerned, anyone preparing the design Presumably it matters not whether the strike is official or unofficial, but it should be noted that a work to rule does not fall within this clause. The strike must operate directly. A strike which affects a trade which in turn affects another trade involved in the Works does not give grounds for an extension unless, of course, it is a transport or manufacturing strike such as is specifically covered. For example, a strike in firm A which manufactures door-closers for the site will qualify, but a strike in firm B which supplies a special component of the door-closers to firm A will not, even though progress may be as badly affected by one as by the other.

The exercise of any statutory power by the UK government, after the date of tender, which directly affects the Works This clause is quite broad in its effects. The events would probably be covered by *force majeure* in any case. The key word is 'directly'. Thus, a government order which set in train a series of events that culminated in such restrictions would not be covered.

Delay in the receipt of permission or approval of any statutory body This applies to planning permission and the satisfaction of building regulations. It also covers such things as entertainment licences and any other applicable statutory control. This applies only to DB.

Force majeure This is a wider term than 'act of God' meaning broadly 'circumstances independent of the will of humankind'. Most situations which would be covered by this clause, such as strikes, war, etc., are already dealt with under other clauses.

Note that under SBC, DB, IC and ICD, fluctuations are frozen at contract completion date or any extended date. However, there is an important proviso that the architect or the employer under DB must respond to every written notice of delay by granting an extension or otherwise, and the printed text of the extension of time clause must not be amended. If the architect does not respond, or amendments are made, fluctuations continue after the contractual completion date unless amendments are also made to the fluctuations clauses.

Whatever the contract may say, the contractor has a better chance of obtaining an extension if it presents a clear, well-documented case. A network analysis is invaluable for this purpose.

Liquidated damages

There is much misunderstanding of the liquidated damage clause. Its purpose is to provide a sum of money to be paid by the contractor for every week (or day) by which the contractor fails to complete the Works after the contract, or extended, date for completion. The sum must be a genuine pre-estimate of the likely loss to the employer. Such damages do not have to be proved. A precondition in the case of SBC, DB, IC and ICD is that the architect has issued the certificate (or under DB the employer has issued a notice) of non-completion in accordance with clause 2.31 (SBC), 2.28 (DB) and 2.22 (IC/ICD). If the architect fixes a new completion date after issuing a certificate of non-completion, the certificate is thereby cancelled and a new certificate is necessary.

It has always been considered that another precondition was that the employer should give written notice of intention to deduct or require payment.[18] It is referred to in SBC clause 2.32.2, DB clause 2.29.2 and IC/ICD clause 2.23.2. It also seems to be implicit in clauses 2.32.4 (SBC), 2.29.4 (DB) and 2.23.3 (IC/ICD), which state that the employer's written requirement remains effective unless withdrawn even if a new certificate or notice of non-completion has to be issued.

It has been suggested that the employer's written notice of intention to deduct may not be a precondition to deduction of liquidated damages and that all that is necessary is that the contractor must be in no doubt that the employer is exercising contractual power to deduct. This quaint line of reasoning suggests that a cheque sent by the employer, in respect of a financial certificate issued by the architect but reduced to take account of liquidated damages, may itself be a sufficient written requirement.[19] The position has been clarified by a case which reaffirms that the employer's written requirement is a precondition.[20]

In the written requirement the employer must state that he or she may require payment or may deduct liquidated damages. The employer must do this before the date of the final certificate. In addition, of course, if the employer intends to deduct, the appropriate notice must be served five days before the final date for payment of a particular certificate (see section 5.1). If, unusually, the employer requires the contractor to pay, a special written notice requiring payment at the rate in the Contract Particulars must be sent to the contractor by the employer and if the contractor fails to pay, the money may be recovered by the employer as a debt. The last date for serving either notice is five days before the final date for payment of the final certificate.

The employer may deduct the damages from money due to the contractor. The employer is entitled to deduct even if the loss is less than the stated sum or if the employer in fact gains by the delay.[21] The liquidated damages clause is often referred to by the contractor as the 'penalty' clause. This is incorrect. A penalty clause is not enforceable. It matters not whether the sum is referred to as a penalty or as liquidated damages.[22] It is the real nature of the sum which counts. A penalty is a punishment. It would be a penalty, for example, if the stipulated sum were greater than the greatest loss which could conceivably follow from the contractor's breach. Another example of a penalty is where the same amount is deductible as damages upon the happening of dissimilar events. In practice, this most often arises because one sum is stated as damages whether the whole or any part of a readily divisible building is delayed.[23]

Where liquidated damages in respect of an estate of houses were expressed as £x per dwelling per week and no Sectional Completion Supplement was used, it has been held to be inconsistent and unenforceable.[24] The insertion of liquidated damages into a contract requires the greatest care. The wording '£*x* for every week or part of a week' is capable of being construed as a penalty, for the obvious reason that damages incurred during a one-day delay are unlikely to be equal to damages during a full week's delay, unless, of course, they represented something like rental charges payable at the beginning of each week. Although it is possible to substitute unliquidated for liquidated damages, it cannot be accomplished simply by inserting 'nil' for the amount of liquidated damages. That would simply mean that the employer has specified 'nil' as the total amount of liquidated damages suffered as a result of delay.[25] It would be necessary to delete the whole of the liquidated damages clause (clause 2.32 in SBC, 2.29 in DB and clause 2.23 in IC/ICD). The courts will not take account of hypothetical situations when considering this question. They will take a pragmatic approach to whether the sum is liquidated damages or a penalty.[26]

In deciding whether a sum is liquidated damages or a penalty, the court would look at the situation at the time the contract was entered into, not at the time of the breach. Provided that the sum is a genuine attempt to estimate the employer's loss in the future, it will be valid. It is of no consequence that it is virtually impossible to estimate accurately, provided that the employer makes reasonable assumptions.

Contractors sometimes say that if there is a liquidated damages clause there must be a corresponding bonus clause for early completion. There is no legal foundation for such statements. It is entirely up to the employer whether he includes a bonus clause. If he does, it need have no monetary relation to the liquidated damages clause. Such a bonus can be as large or small as the employer wishes. Of course, once the sum is inserted in the contract it must be paid if early completion is achieved, or it may form part of a claim if the contractor is delayed by the employer or the architect.

If an employer tells the contractor that liquidated damages will not be deducted and the contractor acts on that basis (such as paying all the sub-contractors in full), the employer will probably be estopped (prevented) from later attempting to recover the liquidated damages.[27]

6.2 Money claims

Under the standard building contracts, to make a claim implies that a contractor considers that it is due an extra payment quite apart from the ordinary system of valuing work done. Claims are a symptom of inefficiency; whether claims are made on the part of the contractor or the architect and employer depends on the circumstances. Contractors generally dislike making claims; architects always dislike dealing with them.

There are three kinds of claim: contractual, common law (sometimes called extra-contractual) and *ex gratia*.

Contractual claims are those claims which are made under the express provisions of the particular contract in use. SBC makes provision mainly in clauses 4.23 and 4.24, but also in clauses 5.10.2 and 3.24. IC and ICD deal with them under clauses 4.17 and 4.18, but also in clause 5.6.2. DB deals with them under clauses 4.19, 4.20 and 5.7.2, but also under supplementary provision S5 if the Contract Particulars entry states that they are to apply. MW and MWD make no provision for contractual claims but does make limited provision for the architect to take direct loss and/or expense into account when valuing variations (see section 4.2) or if regular progress has been affected by the employer's compliance or non-compliance with clause 3.9 (regarding the CDM Regulations). In order to make a contractual claim, the contractor must comply precisely with the terms of the appropriate clause. More details of this are given below.

Common law claims arise outside the express provisions of the contract. They generally relate to breach of implied or express terms of the contract. Thus, such a claim might arise if the employer hindered the contractor's progress on site. Common law claims can also be made in tort, for breach of a collateral contract for *quantum meruit* or sometimes for breach of statutory duty. Some contractors do not realise that it is possible to make a common law claim as an alternative to a contractual claim on the same facts or if the contractor has not complied with the provisions of the contract claims clause and provided, of course, that the facts constitute sufficient grounds. For example, architects' instructions form grounds for a claim under the contract machinery, but not at common law, because the issue of architects' instructions is not a breach of contract; quite the reverse: it is something expressly allowed by the contract. Contractors should take care, however, that all claims are submitted and dealt with before the final certificate is issued if SBC, DB, IC and ICD are involved. The final certificate under those forms is now conclusive that all claims of whatever kind, based on the relevant matters, have been settled.

Ex gratia claims are sometimes known as hardship claims. They have no legal foundation, but the contractor sometimes considers that it has a moral right to extra payment or it has, perhaps, underpriced and it appeals to the employer to help. Whatever the moral rights may be, the contractor's chances with such a claim will usually depend on the benefit to the employer – if any. If to make an *ex gratia* payment would enable the contractor to continue with a contract when otherwise it would go into liquidation, the employer may occasionally make a payment to save the expense of securing another contractor to continue the work at a very much increased cost.

SBC clauses 4.23 and 4.24, DB clauses 4.19 and 4.20 and IC/ICD clauses 4.17 and 4.18 provisions can be considered together because they are very similar in most respects. There is no connection between financial claims – known in the contract as applications for direct loss and/or expense – and claims for extension of time. The contractor must be able to show that the regular progress of the Works has been or is likely to be materially affected by one or more of the list of matters included in the clause. 'Materially' means substantially. Trivial disruptions are excluded. The grounds included are, briefly:

- deferment of possession;
- variations;
- architect's or (under DB) employer's instructions with regard to postponement, discrepancies, provisional sums;
- opening up of work found to be in accordance with the contract;
- delay in receipt of development control requirement permission;
- where approximate quantities are included in the bills of quantities and the approximate quantities are not a reasonably accurate forecast of the quantity of work required (not DB);
- suspension of the contractor's performance of obligations;
- impediment, prevention or default of the employer.

Contractor's duties

In order to be able to claim, the contractor must have suffered or be likely to suffer direct loss and/or expense. This means damages as generally understood. The word 'direct' is important. Direct damage flows naturally from the breach without any other intervening cause.[28] Therefore, if (under SBC, IC or ICD) the architect issues an instruction varying the type of ironmongery and the contractor incurs loss and/or expense over and above the difference in the value of the work and materials, it is entitled to claim. But if the cause of the contractor's loss and/or expense is the fact that the supplier fails to deliver the new ironmongery according to an agreed delivery date, the supplier's failure would be an intervening cause and the contractor could not cite the architect's instruction as giving rise to direct loss and/or expense. Direct damage also means the normal and ordinary damage which does depend upon special circumstances. In order to claim such special damages in addition to the direct damages, the contractor would have to proceed at common law and show that the special circumstances were known to the parties at the date the contract was entered into.[29] The precise circumstances of every case require careful study.

It is vital that the contractor makes its application as soon as it is aware that the regular progress is being or is likely to be affected. Under SBC, IC and ICD, the architect is entitled – indeed, has a duty – to reject late applications. The reason for this provision is clear. The architect must have the opportunity, so far as possible, to issue instructions to overcome the difficulty or to arrange for proper and detailed records to be kept. The architect cannot act without being aware of the difficulty. Therefore, the contractor must not delay making application simply because it has not marshalled all its facts. However, an architect may not be justified in rejecting an application if it can be shown that the fact that it was technically late did not prejudice the employer's position in any way.

The initial application can be quite simple:

> Dear Sir,
>
> We hereby make application under clause 4.23 [substitute '4.19' when using DB or '4.17' when using IC or ICD] of the conditions of contract as follows:
>
> We have incurred/are likely to incur direct loss and/or expense and financing charges in the execution of this contract for which we will not be reimbursed by a payment under any other provision in this contract* because the regular progress of the Works has been/is likely to be materially affected by [describe], being a relevant matter in clause [insert one or more of clauses 4.24.1–4.24.5 when using SBC, clauses 4.20.1–4.20.5 when using DB or clauses 4.18.1–4.18.5 when using IC or ICD].
>
> Yours faithfully,

If deferment of possession is involved, continue as follows from the asterisk: '*owing to deferment of giving possession of the site*'.

Architect's duties

Under SBC, IC and ICD it is for the architect to decide whether the claim is valid. In order for him to be able to form an opinion, further information will usually be requested which the contractor must provide as soon as reasonably possible. If the architect decides that the claim is valid, the amount of direct loss and/or expense must be ascertained by the architect, or the architect may ask the quantity surveyor to ascertain it. Note that it is not up to the quantity surveyor to decide validity; only the architect can do that.[30] The contractor must provide all the information required by the architect or the quantity surveyor to enable them to carry out the ascertainment.

The contractor must make a new application whenever a fresh matter arises. If, however, the matter referred to in the original application is continuing, the architect's duty is to continue to carry out ascertainment, certifying such amounts as soon as each ascertainment is complete. The architect may not simply wait until the end of the contract before certifying such sums. If the architect fails to act, it is a breach for which the employer is responsible.[31]

Loss and/or expense under DB

The contractor may make application for loss and/or expense under clause 4.19. The procedure and the contractual provisions are very similar to SBC, but there are two important differences.

The first and obvious difference is that the application is to be made to the employer, or to the employer's agent if appointed.

The second difference is that whether direct loss and/or expense has been incurred is not a matter of the opinion of the employer or agent, although the employer must be provided with all the information reasonably required in support of the application. Clause 4.19 merely states that the amount of loss and/or expense 'shall' (i.e. must) be added to the contract sum. The reality is that when the contractor makes the applications for payment under clause 4.9, the amount of loss and/or expense which the contractor believes is due will be included. If the employer disagrees with such amount, the remedy is to serve a notice of payment proposal under clause 4.10.3, stating what the employer proposes to pay and how it is calculated, and, if applicable, a withholding notice under clause 4.10.4. The time available to the employer for considering the loss and/or expense and serving the notice is only nine days from receipt of the application. Presumably, the way the clause is intended to work is that the contractor makes the application, the employer calls for reasonable supporting information and only after that does the contractor include the amount in its application. If the contractor attempts to bypass the procedure by including the amount in the application before supplying the supporting information, the employer may simply exclude the amount by issuing a withholding notice on that ground. Clause 4.9.3 gives some assistance to the employer by providing that the contractor must give such details with its application for payment as the employer has requested in the Employer's Requirements. Of course, it is a matter for the employer to ensure that the appropriate details are so requested.

The contract provides for a much simpler process if supplemental provision S5 applies. Then the procedure is that with each application for payment, the contractor must include an estimate of the amount of loss and/or expense incurred during the preceding period. As long as the loss and/or expense continues, the contractor must continue including the estimates. The employer has twenty-one days to accept the estimate, negotiate, refer to adjudication (seems rather extreme at that point) or say that clause 4.19 will apply in the normal way.

If there is agreement, or agreement after negotiation, the amount is to be added to the contract sum. The contractor can claim it in the next payment application.

There are two possible stings in the tail for contractors. The first is that after agreement of an estimate for a particular period, the contractor may not revisit that estimate even if it later discovers that it has overlooked something very serious. It should be remembered, however, that the contractor still has its common law remedies available for damages if the serious matter is a breach of contract.

The second sting is that if the contractor fails to send the estimates, the loss and/or expense is to be dealt with under clause 4.19, but the contractor is not entitled to receive payment until the final account and final statement. Moreover, no financing charges are payable.

Key points

The presentation of the supporting information for a claim requires and deserves careful thought. Some contractors appear to think that if they send large parcels of documents, the architect will be obliged to certify something approaching the sum required. However, so long as the architect is baffled by the claim, no decision will be made. The correct approach is to make a claim as simple and easy to digest as possible. The main points should be logically arranged, based on and referenced to contract clauses. Each point should be cross-referenced to evidence bundled separately. Long and rambling narratives are a mistake. Indeed, they are usually counterproductive.

The contractor who wishes to avoid long delays and numerous requests for further information will submit its supporting information in this way as soon as possible after making the initial application, without waiting for the architect or the employer under DB to request it.

What may be included in a claim? The answer is any direct loss and/or expense directly referable to one or more of the relevant matters in the clause. What constitutes direct loss and/or expense depends upon the particular circumstances. It can become quite complicated, but attention should be given to the following:

- Plant and labour inefficiency as a direct result of the disruption.
- Increases in cost occurring during the period of disruption.
- Increases in head office overheads. This is a difficult cost item to disentangle and many contractors put forward costs based on a formula. The most used is the Hudson formula, but Eichlay and Emden formulas are also used. The Emden formula is probably the most convincing. Before any entitlement to overheads can get off the ground, the contractor must be able to show that other work had to be turned down owing to the prolongation on site. This can be difficult to demonstrate, particularly where large firms are involved.[32]
- Establishment costs calculated at the time of disruption.
- Loss of profit is a permissible part of a claim.[33] The contractor must be able to show that it would have been able to earn the profit elsewhere but for the disruption.[34]
- Interest and financing charges are allowable provided they are a direct result of the disruption.[35]
- Cost of preparing the claim cannot normally be claimed except in so far as it forms part of head office overheads. There may be a case (although difficult to prove) for a contractor to claim, as special damages, the cost of preparing a claim to satisfy an unreasonable architect or quantity surveyor of the justice of a claim.[36]

The key point to be borne in mind is that loss and/or expense can be claimed provided it is a foreseeable and direct result of the disruption or delay. Moreover, although the contractor must prove that it has suffered or incurred the loss before it is entitled to reimbursement, the burden of proof is 'the balance of probabilities'. This is very much less than the standard of 'beyond reasonable doubt' in criminal cases. The balance of probabilities is like saying that it is more likely than not that the contractor suffered the loss.

If there is an extremely complicated interaction between a number of matters included in the claims clause such that it is virtually impossible to separate the costs into a series of items, it is possible to present the evaluation on a global basis.[37] In such cases it is important to ensure that there is no duplication and to identify individual causes where this is possible.

The distinction must be made between global evaluation, which may be permitted in appropriate circumstances, and the global approach to liability, which is not permitted.[38] A contractor must frame its application with sufficient detail that the architect is able to see exactly to what matters the contractor is referring and in what way each of the matters delayed or disrupted the Works. For example, it is not good enough for the contractor simply to say, 'There were a multitude of variations throughout the project, AIs [architect's instructions] were late, the employer's men were constantly tramping through the site and the result was an overrun of six weeks and £24,000 extra loss and expense.' Instead, the contractor should list each variation and when it was received, and demonstrate the effect of each one individually, and it should do the same with the late AIs and the occasions when the employer's men disrupted the Works. If the claim is for costs due to delay, and the consequences of numerous events have a complex inter-action, it may be permissible to maintain a composite claim if it is impossible to identify the specific relation of each event and its time and cost result.[39] The circumstances when this approach will be acceptable will be rare in practice. A contractor is entitled to submit a global claim if it wishes, but it should be aware that there are serious difficulties of proof.[40] Global claims, if presented as 'all or nothing', will fail completely if some of the causative events are not established. Therefore, this type of claim should show how the total amount is to be reduced if some events are not established and to take account of the contractor's own inefficiencies and issues of that nature.[41]

It often happens that a contractor suffers loss and/or expense due to a variety of causes, some of which are permissible as contractual claims, others of which fall outside the claims clause and can form the basis only of a claim at common law. Strictly, common law claims must be taken through the courts, but it usually does no harm to submit them to the architect as part of an overall claim provided it is made clear that they are not contractual claims. The architect should refer all common law claims to the employer, who might well decide that it is more convenient to settle them through the architect than through arbitration or the courts. Of course, any money found to be due to the contractor as a result of a common law claim cannot be certified through the contract. It must be paid by the employer separately. All claims for loss and/or expense under MW and MWD, with the very limited exception already noted, are common law claims, because there is no contractual machinery to enable the contractor to claim and the architect to deal with such claims. Finally, it should be noted that whether the contractor is claiming through the contract or at common law, it can only claim – and the architect can only ascertain – the precise loss and expense suffered.[42]

References

1 Dodd v Churton (1897) 1 QB 562.
2 Wells v Army & Navy Co-operative Society Ltd (1902) 86 LT 764.
3 Percy Bilton Ltd v Greater London Council (1981) 20 BLR 1.
4 Peak Construction (Liverpool) Ltd v McKinney Foundations Ltd (1970) 1 BLR 111.
5 Scott Lithgow v Secretary of State for Defence (1989) 45 BLR 1.
6 H. Fairweather & Co. Ltd v London Borough of Wandsworth (1987) 39 BLR 106; Methodist Homes Housing Association Ltd v Messrs Scott & McIntosh 2 May 1997 unreported.
7 Ascon v Alfred McAlpine (1999) CILL 1583.
8 London Borough of Merton v Stanley Hugh Leach Ltd (1985) 32 BLR 51.
9 John H. Haley Ltd v Dumfries and Galloway Regional Council (1988) GWD 39–1599.
10 Victor Stanley Hawkins v Pender Bros Pty Ltd (1994) 10 BCL III.
11 Alghussein Establishment v Eton College [1988] 1 WLR 587.
12 Amalgamated Building Contractors Ltd v Waltham Holy Cross UDC [1952] 2 All ER 452.
13 Temloc Ltd v Errill Properties Ltd (1987) 39 BLR 30.
14 Cantrell and Another v Wright & Fuller Ltd (2003) 91 Con LR 97.
15 Rapid Building Group Ltd v Ealing Family Housing Association Ltd (1984) 1 Con LR 1.
16 Walter Lawrence & Son Ltd v Commercial Union Properties (UK) Ltd (1984) 4 Con LR 37.
17 Levy v Assicurazioni Generali [1940] AC 791.
18 A. Bell & Son (Paddington) Ltd v CBF Residential Care and Housing Association (1989) 5 Const LJ 194.
19 Jarvis Brent Ltd v Rowlinson Constructions Ltd (1990) 6 Const LJ 292.
20 Finnegan v Community Housing (1993) 65 BLR 103.
21 Clydebank Engineering v Yzquierdo Castenada [1905] AC 6.
22 Dunlop Pneumatic Tyre Co. Ltd v New Garage Motor Co. Ltd [1915] AC 79.
23 Stanor Electric Ltd v R. Mansell Ltd (1987) CILL 399.
24 Bramall & Ogden Ltd v Sheffield City Council (1983) 1 Con LR 30.
25 Temloc Ltd v Errill Properties Ltd (1987) 39 BLR 30.
26 Phillips Hong Kong Ltd v Attorney General of Hong Kong (1993) 9 Const LJ 202.
27 London Borough of Lewisham v Shepherd Hill Civil Engineering 30 July 2001 unreported.
28 Croudace Construction Ltd v Cawoods Concrete Products Ltd (1978) 8 BLR 20.
29 Hadley v Baxendale (1954) 9 LR Ex 341.

30 John Laing Construction Ltd v County & District Properties Ltd (1982) QB 24 November.
31 Croudace Ltd v London Borough of Lambeth (1986) 6 Con LR 70.
32 AMEC Building Contracts Ltd v Cadmus Investments Co. Ltd (1997) 13 Const LJ 50.
33 Saint Line Ltd v Richardsons Westgarth & Co. Ltd [1940] 2 KB 99.
34 Peak Construction (Liverpool) Ltd v McKinney Foundations Ltd (1970) 1 BLR 111; City Axis v Daniel P. Jackson (1998) CILL 1382.
35 Rees & Kirby Ltd v Swansea City Council (1985) 5 Con LR 34.
36 Babcock Energy Ltd v Lodge Sturtevant Ltd (formerly Peabody Sturtevant Ltd) (1994) 41 Con LR 45.
37 J. Crosby & Sons Ltd v Portland UDC (1967) 20 BLR 34.
38 Wharfe Properties Ltd v Eric Cumine Associates (1991) 52 BLR 1.
39 Mid Glamorgan County Council v J. Devonald Williams & Partner (1991) 8 Const LJ 61.
40 GMTC Tools & Equipment Ltd v Yuasa Warwick Machinery Ltd (1994) 73 BLR 102.
41 How Engineering Services Ltd v Lindner Ceiling Partitions plc 17 May 1995 unreported.
42 Alfred McAlpine Homes North Ltd v Property & Land Contractors Ltd (1995) 76 BLR 65, but see also How Engineering Services Ltd v Lindner Ceiling Partitions plc [1999] 2 All ER 374.

7 The end

7.1 Practical completion and defects liability

When the architect is of the opinion that practical completion has been achieved and the contractor has provided all the information necessary for the health and safety file, the architect must issue a certificate to that effect, stating the date of practical completion (SBC clause 2.30, IC/ICD clause 2.21, MW clause 2.9 and MWD clause 2.10). Under DB, there is no certificate. When the Works have reached practical completion and the information for the health and safety file has been provided by the contractor, the employer must issue a written statement to that effect. It is treated as a matter of fact, not opinion.

The contractor is always anxious to obtain this certificate, which confers enormous benefits. It marks the date when:

- The rectification period begins.
- The contractor's liability for insurance ends.
- The employer's right to deduct full retention ends and half the retention already held is due for release.
- Liability for liquidated damages ends.
- The contractor's licence to be on the site ends.
- Regular interim certificates cease.
- The period of the review of extensions of time begins (not applicable to MW and MWD).

Such is the importance of practical completion that the contractor might be forgiven for thinking that the term would be carefully defined in the contract. Unfortunately, no definition is to be found. This does not mean that the architect, or the employer under DB, is completely free to fix any date. Except under DB, the date is left for the architect's opinion, but that opinion is open to question if the contractor chooses to refer the point to adjudication or arbitration. A number of cases have considered the matter, not always coming to quite the same conclusion. However, a number of important factors can be identified. Practical completion does not mean substantially or nearly complete. It means complete except for minor items when there are no apparent (i.e. visible) defects.[1] It certainly does not mean completion down to the last detail. That would make it a penalty clause and, therefore, unenforceable.[2] A working definition might be that practical completion is when there are no defects apparent and when such minor items as are left to be completed can be completed without any inconvenience to the employer using the building as intended. A contractor who feels that the practical completion certificate or written statement is being unreasonably withheld should inform the architect or the employer under DB in writing, noting the date on which the contractor considers practical completion has been achieved.

On many projects it is common to have what is known as a 'handover meeting', at which the employer and the professional team meet the contractor and jointly inspect the building. In some instances the decision that the building has reached practical completion is effectively left to the employer. Such an arrangement is quite wrong, and if the employer were to make the final decision it would be a breach of the contract, which stipulates (except under DB) that the decision is one for the architect alone. If the employer interferes with or obstructs the issue of any certificate, the contractor has grounds for termination (SBC, IC and ICD clause 8.9.1.2, MW and MWD clause 6.8.1.2).

Whether there would be much point in termination at such a late stage in the contract is a matter to be decided on the particular circumstances. It is probably more usual that the employer is anxious to accept the building, often against the architect's advice. Since practical completion is greatly to the contractor's advantage, interference by the employer to secure the early issue of the certificate would seldom be questioned by the contractor.

Sometimes, under SBC, IC, ICD, MW or MWD, the employer will take possession of the building some time before the architect would have certified practical completion. The contractor does not dissent, and both employer and contractor expect the architect to issue a certificate of practical completion. The architect may consider that, if both parties agree, he or she is free to issue the certificate. This is not, in fact, the case: the contract requires the architect to issue the certificate only when certain criteria have been satisfied. The certificate is the formal expression of the architect's opinion.[3] If the architect's opinion was not properly represented in the certificate, it would amount to unprofessional conduct, to say the least. Quite apart from matters of professional ethics, there could be serious legal consequences for the architect, depending on particular circumstances, despite the agreement of the parties to the contract.

In some instances the decision seems to be taken by the clerk of works. Although this appears to be wrong, it must be remembered that, in forming an opinion, the architect is entitled to take into account any information, including the views of the clerk of works. The final opinion, however, must be that of the architect. A practical completion certificate signed by the clerk of works is simply a useless piece of paper.

Much more serious are the situations where the contractor suspects that a certificate of practical completion is being delayed because the employer/ developer has been unsuccessful in leasing the property, and liquidated damages provide an income until a lease can be arranged. The contractor should be able to rely upon the architect to act properly in accordance with the contract even though considerable pressure may be imposed by the employer.

Under DB, although practical completion is treated as a matter of fact and not opinion, in practice it is for the employer to make the decision to issue the written statement. The contractor's recourse is to adjudication or arbitration.

Where sectional completion is used, a section completion certificate must be issued on practical completion of each section. When the last section completion certificate is issued, a practical completion certificate for the Works should be issued at the same time. That is because the contractor's obligation is to carry out and complete the Works. It is possible that the sum of all the sections may not quite be the same as the whole Works for various reasons. The employer must do the same under DB, using the section completion statements and the practical completion statement for the whole of the Works.

Defects liability

Once practical completion has been certified, the rectification period begins. The length of the period can be anything, provided it is clearly stated in the contract. A common period is six months, but many consider that twelve months is more appropriate, and mechanical installations commonly specify twelve months on all contracts. It should be noted that none of the six contracts under consideration allows for differing periods to be inserted for different parts of the same building unless completion in sections is employed. Problems can also occur if the sub-contract and main contract periods do not correspond.

The rectification period is often mistakenly referred to as the 'maintenance period'. Maintenance implies that the building must be kept in pristine condition: floors polished, paintwork renewed, etc. Contractors sometimes think that they have a duty to attend to any matter raised during this period. That is not the case. The contractor must make good any defects, shrinkages and other faults which appear during the period (SBC clause 2.38, DB clause 2.35, IC/ICD clause 2.30, MW clause 2.10 and MWD clause 2.11). The reference to 'excessive' shrinkages in the 1998 edition of the Agreement for Minor Building Works has now been abandoned.

The phrase 'other faults' is to be interpreted to mean faults which are like defects and shrinkages. 'Defects' means work not in accordance with the contract and not defects due to some other reason, such as poor standard of specification or the employer's occupancy. 'Shrinkages' are to be made good only if they are due to materials or workmanship not being in accordance with the contract. Thus, if the moisture content of the timber was specified too high, but the timber was fixed at the specified moisture content and shrinkage took place solely because, on occupation, central heating reduced the moisture content below the specified level, the contractor would not be liable.

Although it is no longer expressly stated, the contractor is also liable for frost damage, provided the frost occurred before the date of practical completion. Damage due to frost occurring after practical completion is the employer's liability.

Reference is made to defects which 'appear' during the period. The word seems to extend to mean defects which are apparent during the period; for example, defects still existing at practical completion, even though there should be no such defects.[4] In any event, the contractor will be liable for any disconformity between the Works and the contract documents.

If the architect instructs how the defects are to be made good, the contractor may be entitled to payment as though the architect had issued a variation.[5]

MW and MWD are not specific, but SBC, IC and ICD empower the architect to issue instructions that defects are to be made good at any time during the period and up to fourteen days thereafter. The employer is given a similar power under DB. The contractor is not entitled to wait until the end of the period before it commences the making-good process. SBC and DB require the architect (the employer under DB) to specify the defects in a schedule which must be delivered within fourteen days after the end of the rectification period. No further instructions may be issued under the defects clause after the issue of the schedule or after the end of the fourteen days, whichever is earlier.

The significance of this is sometimes misunderstood by contractors. The contractor can refuse to make good defects after the end of the period, but if work or materials are not in accordance with the contract, they are still the contractor's liability. The rectification period is largely for the contractor's benefit. If it were not for this period, the contractor would have no right to return to the site and rectify defects. The employer could simply take adjudication or arbitration proceedings for damages (probably the cost of having the work put right by others). This is the position after the end of the period. In general, any defects which appear after the rectification period will be referred to the contractor, in the hope of obtaining speedy rectification. If the contractor refuses, and the defects are a result of the work or materials not being in accordance with the contract, the employer can employ others and adjudicate or arbitrate against the contractor. The employer has the right to charge the contractor the full cost of putting right defects which appear after the end of the rectification period.

The contractor is to make good the defects at its own cost. The architect or the employer under DB is empowered, however, to instruct that some or all of the defects are not to be made good (with the consent of the employer), in which case an appropriate adjustment may be made to the contract sum. The precise meaning of 'appropriate deduction' has been the subject of some debate. It is considered that it refers to something less than the cost of having the work done by others. Such an instruction may be issued if the employer does not want to be disturbed by the rectification of minor faults or because the architect considers that it would be prudent to employ someone else to do the work. The defects clause gives the contractor the right to reduce the cost of remedial works by doing them itself.

It is now established that where, without fault on the part of the contractor, the employer elects not to have the defects corrected by the contractor, the sum that can be recovered (appropriate adjustment) is limited to the sum which represents the cost which the contractor would have incurred if it had been called upon to remedy the defects.[6] It appears that if defects which appear during the rectification period are not notified to the contractor until later, it is still a breach of contract on the part of the contractor, but the contractor cannot be made to put the work right and the employer can recover from the contractor only what it would have cost the contractor to correct the defects.[7] If some of the defects are the responsibility of the employer, the contractor may be instructed that such defects are to be made good but that the contractor is to be paid for so doing. It is not thought that the contractor is obliged to deal with such defects even if paid. At the end of the rectification period, when all defects have been made good, the architect must issue a certificate to that effect. Under SBC terms the second half of the retention will be released after the issue of this certificate. Under DB the release of the second half of the retention is triggered by a notice of completion of making good defects.

Partial possession

MW and MWD contain no provision for the employer to take partial possession of the Works. SBC clauses 2.33–2.37, IC/ICD clauses 2.25–2.29 and DB clauses 2.30–2.34 refer to partial possession. It is important to remember that partial possession is not the same as phased or sectional completion. If the employer wishes the Works to be completed according to a particular timetable, sectional completion, which is designed for this purpose, should be used and the appropriate parts of the Contract Particulars should be completed. If the standard form is used without sectional possession and completion, the employer cannot achieve a phased completion by simply – as is sometimes done – setting out the phases and completion dates in the bills of quantities. The contractor's obligation in such a case is simply to complete by the date for completion in the Contract Particulars. Any conflicting dates in the bills are overridden by the provisions of SBC, IC and ICD clause 1.3. Similarly, conflicting dates in the Employer's Requirements are overridden by clause 1.3 of DB.

The purpose of the partial possession clause is to enable the employer, with the consent of the contractor, to take possession of any part or parts of the Works before practical completion of the whole of the Works. If, unusually, partial possession was to be taken of the whole of the Works, it would amount to practical completion.[8] Note that the employer cannot take possession without the contractor's consent (the contractor is in possession of the site), but the contractor may not refuse such consent unreasonably. If the contractor agrees to the employer taking possession, the contractor should take care that the procedure is carried out formally. It should give consent in writing subject to a written statement being issued stating the date on which partial possession was taken in accordance with clause 2.33 (IC/ICD clause 2.25 or DB clause 2.30). The statement must be issued as soon as partial possession is taken, but the contractor would be advised to secure the statement no later than the date of possession and make the handing over of the written statement a condition for the giving of consent. The reason for taking such great care is that it has been held that if the employer merely enters the Works, stores materials and even carries out factory processes in a building before practical completion, but without the issue of any formal certificate or written statement, partial possession has not been taken. Therefore, if there is a fire which destroys the building and the contents brought in by the employer, the contractor is liable for the cost unless the position has been agreed with the insurers.[9]

The issue of the written statement, however, will ensure that the appropriate contractual provisions come into operation as follows:

- The rectification period is deemed to have begun, for the part taken into possession, on the date in the written statement.
- When all defects, etc. have been made good, a making-good certificate (or notice in the case of DB) for the part is to be issued.
- The contractor's obligation to insure the part ceases (this is a crucial provision, because the responsibility becomes the employer's).
- The amount of liquidated damages which may be payable is reduced in proportion to the value of the part.

Provided that the contractor is not inconvenienced by the partial possession of the Works, and it secures the written statement, it can be seen that there are advantages in acceding to the employer's request.

SBC, IC and ICD clause 2.6 and DB clause 2.5 contain provisions to cover the situation where the employer wishes to make use of or occupy the Works before practical completion for storage or other purposes, but does not wish to take partial possession. The contractor's consent is required, but the contractor is not to withhold such consent unreasonably. Before the contractor gives its consent, the party responsible for Works insurance must obtain from the insurers confirmation that the proposed use will not prejudice the insurance. If the contractor is responsible for insurance and the insurers require an additional premium, the contractor must notify the employer, and if the employer still requires the use of the Works, the amount of the additional premium payable by the contractor must be added to the contract sum. It should be noted that occupation of the Works in this way is not partial possession, and the consequences of partial possession (reduction of liquidated damages, start of the rectification period, etc.) do not follow.[10]

7.2 Suspension and termination

If the employer fails to pay the contractor in full the amount due by the final date for payment, the contractor may issue a notice to the employer stating its intention to suspend. It must also state the grounds of the suspension (i.e. that the employer has failed to pay a particular sum of money which should have been paid by a specified date). The notice must be written and it must allow the employer seven days in which to pay the money properly due. Obviously, the employer is not obliged to pay any amount for which a valid notice has been issued saying that it is intended to withhold payment. It should be noted that under SBC, IC, ICD, MW and MWD, in order to comply with the clause the contractor must send a copy of the notice to the architect. This is not a requirement of the Housing Grants, Construction and Regeneration Act 1996, and if the contractor omitted to send a copy to the architect, it would not stop it from being entitled to suspend work.

This is a draconian remedy, but no doubt merited on occasion. It is suggested that the notice should be sent by special delivery to ensure that it is received the next business day and that there is evidence to that effect. The suspension may continue until payment has been made in full. The clause goes on to say, reflecting the relevant section of the Act, that the suspension should not be treated as a suspension to which the termination clause refers, neither is it to be treated as a failure to proceed regularly and diligently. That should go without saying. The clause is said to be without prejudice to any other rights and remedies which the contractor may possess. This means that if the contractor exercises this right, it does not prevent it from exercising other rights under the contract or at common law.

All JCT contracts provide the contractor with further remedies to deal with the period of suspension. It is to be taken into account for an extension of time, and the contractor is also entitled to make applications for loss and/or expense except under MW and MWD, where the contractor must seek common law remedies.

Under the general law it is possible to discharge a JCT contract in one of four ways:

- By performance, when both parties have completed their obligations under the contract.

- By agreement, when both parties agree that, despite not having completed or perhaps started their obligations to each other, the contract should come to an end. This should be done with the same formality with which the original contract was entered into. Unless executed as a deed, such an agreement will usually be effective only if both parties have outstanding obligations.
- By frustration, when events outside the control of both parties render the contract something radically different from what was originally contemplated.[11]
- By breach, when one of the parties commits a breach so serious as to indicate a clear intention not to be bound by the contract and the other party accepts that the contract is at an end. There are serious dangers for the party accepting the repudiation, which will be discussed later.

All four contracts state that the contractual right to terminate is without prejudice to any other rights and remedies which the employer or contractor may possess. In other words, the employer or the contractor may opt to bring their obligations to an end, if they have sufficient grounds, using common law principles rather than the contractual machinery. In certain instances it may pay them to do so because termination at common law entitles the party properly carrying out the termination to damages, whereas termination under the contract simply entitles that party to whatever remedies are stipulated in the contract.[12] However, it should be noted that some of the grounds listed in the contract entitling the employer or contractor to terminate the contractor's employment would not be sufficient to justify termination at common law.

The four contracts refer to termination of employment; thus, the contract itself is still in existence and provides for the arrangements to be made after termination. If termination of the contract takes place at common law, the whole contract is extinguished and none of its clauses can apply thereafter except the dispute resolution clauses. This is rare, and unless the contract or notice is specific, a court will consider that the parties intend merely to put an end to their remaining obligations under the contract.[13]

The termination clause in SBC, IC, ICD and DB is clause 8. There is an additional clause related to insured risks in SBC and DB schedule 3 paragraph C4.4, and in IC and ICD it is schedule 1 paragraph C4.4 and provision for termination by the employer if terrorism cover is to cease (SBC, DB, IC and ICD clause 6.10.2.2). In MW and MWD, termination is covered by clause 6.

Damage by insured risks

SBC and DB schedule 3 paragraph C4.4 and IC/ICD schedule 1 paragraph C4.4 may be considered together because they are identical. They deal with the situation where loss or damage is caused to the Works by one or more of the insured risks if the insurance is the responsibility of the employer and the Works are alterations or extensions to an existing building.

The contractor must give written notice to the employer as soon as it discovers the loss or damage, irrespective of whether it intends to implement the termination provision. The contract states that 'if it is just and equitable' to do so, either party may serve notice on the other to terminate the contractor's employment. The notice must be served by actual, special or recorded delivery within twenty-eight days of the occurrence and then the other party has seven days within which to refer to the dispute resolution procedures as to whether the termination is 'just and equitable'. If notice to refer is not given within the seven days, the right is lost.

Three points deserve expansion:

- The meaning of 'just and equitable' is not clear. The provision is presumably intended to cover the position where alterations are being carried out to an existing building which is itself badly damaged by the occurrence. In such circumstances it would seem pointless to proceed with a contract for alterations if the building to be altered is largely destroyed. Such an event would probably rank as frustration under the general law in any event.

- Where a notice is to be served by actual, special or recorded delivery, it is wise to follow the direction precisely. Simply sending the notice by fax, e-mail or even ordinary first-class post may not suffice. When either party takes a serious step under the contract, it is safest to comply with the provisions exactly, to the extent of repeating the wording in the contract in any notice required to be given, so that there can be no doubt about intentions. The courts tend to take a businesslike view of such items as notices, and may overlook minor imperfections, looking at the spirit rather than the letter, unless it is thought that one of the parties is seeking unfair advantage, when a notice may be interpreted very strictly.

- The termination notice must be given within twenty-eight days of the occurrence itself, not the discovery of the occurrence. Although it seems unlikely that damage of sufficient seriousness to warrant termination under this clause will lie undiscovered for more than a day or so, it is quite conceivable that by the time the contractor has found the damage and notified the employer, the employer may have only three weeks in which to assess the problem and decide what to do.

The consequences of termination under this provision are the same as for termination by either party. The contractor is not entitled to receive any payment in respect of direct loss or damage arising from the termination.

Termination by employer

SBC, DB, IC and ICD clauses 8.4–8.6 deal with termination by the employer. They are similar and can be considered together. The equivalent provision in MW and MWD (clause 6.4–6.6) will be discussed separately.

The grounds for termination are if the contractor:

- wholly or substantially suspends the carrying out of the work before completion without reasonable cause;
- fails to proceed regularly and diligently with the Works or with the design under DB, ICD or MWD where the CDP applies;
- refuses or neglects to remove defective work, etc. after written notice and the Works are materially affected;
- fails to comply with clauses restricting sub-letting, restricting assignment or dealing with named sub-contractors (IC/ICD only);
- fails to comply with the CDM Regulations;
- becomes insolvent or makes an arrangement with creditors, etc.;
- is guilty of corruption, i.e. giving or receiving bribes and the like.

In practice, there may be instances where it is difficult to show that the contractor has wholly suspended without reasonable cause or failed to work regularly and diligently. The employer seeking to rely upon these grounds must be very sure of the facts. It has been suggested that a term in a contract requiring a contractor to proceed with due diligence is an obligation to execute the Works so that key dates and the completion date will be met.[14] Since most contracts, in the absence of sectional completion provision, will not have key dates other than the completion date, it seems on that basis that the contractor's contention that it is working regularly and diligently will be difficult to disprove unless it is clear that it cannot meet the completion date. Architects who fail to take action in respect of a contractor who fails to proceed regularly and diligently may risk legal action by the employer.

'Regularly and diligently' has been considered by the courts and it has been stated that it means 'essentially to proceed continuously, industriously and efficiently with appropriate physical resources so as to progress the Works steadily towards completion substantially in accordance with the contractual requirements as to time, sequence and quality of work'.[15] Certainly, it is not enough if the contractor merely fails to work according to programme or stops working on part of Works for a brief period.

It is difficult to understand why the third ground is included at all. The employer already has a satisfactory remedy if the contractor fails to comply with a notice requiring compliance with an instruction (see section 4.2). In order for this ground to be effective, the contractor must refuse or neglect to deal with defective work. That is, it must fail to act despite reminders. Moreover, its inaction must result in the Works being materially affected – presumably if the defective work would have an adverse effect upon what followed. It is not thought that failure to remedy defects which are not urgent would come within this ground. The Amendments to JCT 80, CD 81 and IFC 84 in the late 1980s removed the requirement that the neglect must be persistent. The argument was apparently that the proviso that termination should not be unreasonable or vexatious would prevent the employer terminating if the contractor neglected to comply with an instruction on just one occasion. This argument appears to be flawed, because if compliance with the instruction was sufficiently important, it is quite possible for termination to be reasonable after one or two instances of neglecting to comply on the part of the contractor. This is a point to be particularly aware of.

The fourth ground is intended to give the employer a remedy if the contractor does not comply with the clauses noted. It is thought that the contractor's failure would have to be quite unequivocal. In most cases the failure could be dealt with less drastically by an appropriately worded letter.

The fifth ground was introduced to give the employer a remedy if the contractor fails to carry out its contractual duties in regard to the CDM Regulations. Presumably, a grave lapse is envisaged before the provision would be applied.

The procedures in regard to termination following the contractor's insolvency and some other matters have been changed. It is convenient to deal with them together.

Termination is no longer automatic. The definition of insolvency is set out conveniently in clause 8.1. Insolvency is covered in clause 8.5. The employer may terminate by notice at any time. The contractor must inform the employer immediately if it becomes involved in any of the matters set out in clause 8.1. Very importantly, from the date the contractor becomes insolvent, even if the employer has not yet given termination notice:

• Most of the contractual consequences of termination will apply.
• Clauses of the contract which require further payment or release of retention will not apply.
• The contractor's obligations to carry out and complete the Works and the design (in the case of DB, ICD or SBC where CPD applies) will be suspended.
• The employer may take reasonable measures for the protection of the site, the work and site materials. The contractor is expressly required to 'allow and not hinder' such measures.

The employer may terminate on grounds of corruption by simply serving a notice of termination stating the ground.

In order to terminate on any of the first five grounds, the architect, or the employer under DB, must first serve a notice by special or recorded delivery specifying the default. Notice may be served also by actual delivery (note that service by fax is not included). The contractor has fourteen days from receipt of the notice in which to remedy or start to remedy the default. If it takes no action, the employer may then serve notice of termination by actual, special or recorded delivery. The employer has ten days in which to act. Note that the architect is not empowered to serve the actual notice of termination. There is a sting in the tail for the contractor because, if it remedies the default after receiving the default notice, and some time later repeats the default, the employer is entitled to serve notice of termination without first serving a further default notice. To what extent the repeated default must be exactly the same as the original default is a matter of speculation, but it is wise to avoid becoming a test case.

There is an important proviso to the effect that notice of termination must not be given unreasonably or vexatiously. Thus, it would be unreasonable to attempt to take unfair advantage[16] and vexatious to serve notice in an attempt to annoy.

Under MW and MWD, clauses 6.4 and 6.5 entitle the employer to serve notice of termination on four of the grounds already mentioned, i.e. wholly suspending the Works, failure to work regularly and diligently, failure to comply with the CDM Regulations and insolvency of the contractor. The earlier comments on these grounds apply equally to MW and MWD. A default notice is specified to be given by the architect in respect of the first three grounds. The contractor has seven days from receipt to end the default, failing which the employer may terminate by giving a further notice. In the case of insolvency, no default notice is required and the insolvency provisions closely follow those of the other forms. In view of the employer's difficulties in terminating on these grounds, mentioned earlier, a contractor might be forgiven for thinking that, short of corruption or insolvency, it is safe from termination provided that it stays on site and does at least some work. Such a view may be unduly optimistic in the light of current law. The various grounds for termination are briefly summarised in Table 7.1.

Termination by contractor

Failure to pay is a common ground to all four contracts under discussion. This is a valuable option for the contractor. At common law, failure to pay is not usually ground for repudiation, but there are exceptions;[17] the remedy is to use the dispute resolution procedures.

Before issuing the termination notice, the contractor must send a notice specifying the default to the employer. It is a fourteen-day notice under SBC, DB, IC and ICD (clause 8.9.3) and a seven-day notice under MW and MWD (clause 6.8.3). All notices must be given by actual, special or recorded delivery and the remarks on delivery in relation to employer termination are applicable.

Under SBC, IC, ICD, MW and MWD the employer's obstruction of the issue of any certificate (not just financial certificates) is a ground for termination. It is usually difficult to prove that the employer has interfered with the issue of a certificate unless the architect is ill-advised enough to say so in

Table 7.1 Grounds for termination under SBC, IC/ICD, MW/MWD and DB

Ground	SBC	IC/ICD	MW	MWD	DB
By employer					
Contractor wholly or substantially suspends work	X	X	X	X	X
Contractor wholly or substantially suspends design				X	
Contractor fails to proceed regularly and diligently with the work	X	X	X	X	X
Contractor fails to proceed regularly and diligently with the design				X	X
Contractor neglects defective work	X	X			X
Contractor wrongly assigns	X	X			X
Contractor wrongly sub-lets	X	X			X
Contractor fails to comply with named sub-contractor clause		X			
Contractor fails to comply with CDM	X	X	X	X	X
Regulations:					
Contractor becomes insolvent	X	X	X	X	X
Contractor guilty of corruption	X	X	X	X	X
If notice that terrorism cover not available	X	X			X
By contractor					
Employer fails to pay, including VAT	X	X	X	X	X
Employer obstructs certificate	X	X	X	X	
Employer wrongly assigns	X	X			X
Employer fails to comply with CDM	X	X	X	X	X
Regulations:					
Works suspended for specified period owing to:					
Specific instructions	X	X	X	X	
Impediment, prevention or default	X		X	X	X
Employer becomes insolvent	X	X	X	X	X
By either party					
Works suspended for specified period owing to:					
Force majeure	X	X	X	X	X
Instructions to default of statutory undertaker	X	X	X	X	X
Damage due to specified perils	X	X	X	X	X
Civil commotion or terrorism	X	X	X	X	X
Exercise of statutory power	X	X	X	X	X
If just and equitable following insured damage	X	X			X

a letter refusing issue. It would be sufficient if the contractor could show that the employer had instructed the architect to reduce the amount on a financial certificate or requested the architect to delay the issue of, say, a certificate of practical completion.[18] Failure by the employer to comply with the contractual provisions regarding the CDM Regulations is a common ground of all six contracts.

If the Works are suspended for a period of one month (in the case of MW and MWD) or for the period inserted in the Contract Particulars (in the case of SBC, IC, ICD and DB) for any of the reasons listed, the contractor can terminate. Most of the reasons can be readily identified from Table 7.1 and read in detail in the particular contract form. One or two points, however, are worth noting. Prior notice of default must be given. The instructions are given in regard to:

- correction of discrepancies;
- variations;
- postponement of the Works.

In order to be able to terminate on the ground of late instructions, the contractor must be able to show that the architect failed to comply with clause 2.11 or 2.12 (SBC) or clauses 2.10 or 2.11 (IC/ICD). Under DB there is no express reference to instructions.

If the employer becomes insolvent and becomes bankrupt or makes an arrangement with creditors or, being a company, starts liquidation or a number of other matters associated with insolvency, the contractor may terminate its employment.

Termination by either party

There are a number of grounds in SBC, DB, IC, ICD, MW and MWD enabling either employer or contractor to terminate. Damage to Works or existing property has been dealt with.

Under SBC, DB, IC and ICD clause 8.11 and MW/MWD clause 6.10, either party may terminate if the Works are suspended for the period stated in the Contract Particulars (or one month in the case of MW and MWD) by *force majeure*, damage to the Works by one or more of the specified perils; instructions issued in regard to the correction of discrepancies, variations or postponement of work, provided they resulted from negligence or default of a statutory undertaker; civil commotion or terrorism; exercise of statutory power by the UK government directly affecting the Works. DB also includes in the three months' suspension section, delay in receipt of development control permissions.

SBC, DB, IC and ICD clause 6.10.2.2 provides that the employer may give notice of termination if the insurers in the joint names policy give notice that terrorism cover will cease from a specific date (the 'cessation date'). The termination notice must state a termination date, which must be after the cessation date.

Consequences

The consequences of termination vary, depending upon the party carrying out the termination and the clause under which the termination takes place. If the employer terminates under SBC, DB, IC or ICD clauses 8.4–8.6, the rights and duties of the parties may be summarised as follows:

- The employer may take and the contractor must give up possession of the site. After termination, the contractor no longer has a licence to remain on site, and if it does so, it is trespassing. It used to be thought that if the contractor disputed termination it had the right to stay in possession until the termination was shown to be valid.[19] Currently, the contrary and more sensible view prevails that the contractor must give up possession pending arbitration, and if the termination is found to be invalid, the contractor will have an adequate remedy in damages. If the contractor refused to give up possession of the site, it is thought that an English court would grant an injunction to force the contractor to give up possession.[20] All six contracts clearly state that termination by the employer and the consequences are without prejudice to the employer's other rights and remedies. Thus, the contractor must leave the site and await the results of any proceedings.
- The employer may employ and pay others to complete the Works.
- The employer may use any of the contractor's site equipment and site materials to complete the Works but not equipment belonging to others without permission.
- If so instructed, the contractor must remove its temporary buildings, plant, equipment, goods and materials and ensure that other owners of equipment do the same.
- To the extent that they are assignable, and to the extent that it is lawful, the contractor must assign all benefits in agreements with sub-contractors and suppliers to the employer, if the architect (or the employer, under DB) so requires within fourteen days of the date of termination. The employer is entitled to pay these firms and deduct such payments from money due to the contractor.
- Clauses requiring payment or release of retention to the contractor cease to apply. Therefore, the employer need make no further payments to the contractor, even on certificates already issued, until the Works are complete and defects rectified.
- If there is a CDP, the contractor must provide the employer with two copies of all the design documents which have been prepared.
- A reasonable time after the Works are complete and the defects rectified, an account must be drawn up stating:

- the amount of loss and/or expense caused to the employer by the termination, including the cost of having the Works completed by others;
- the amount already paid to the contractor;
- the amount which would have been payable for the Works in accordance with the contract.

The result may be a sum payable to the contractor, but it is more likely to be a sum due to the employer. In effect, the employer is entitled to the difference in cost between the original contract and the actual cost. Proper allowance must be made for variations, professional fees and any other costs incurred by the employer.

In all these contracts (except MW and MWD) the power has been intro-duced for the employer to opt not to continue with the work at all, in which case the contractor is entitled to payment as noted above. To prevent abuse of this power, if the employer does not opt to continue or abandon the Works within six months of the date of termination, the contractor may serve notice on the employer requiring a decision.

In exercising the option of not completing the Works, the employer must notify the contractor within six months of termination. The notice must be followed with a statement setting out the value of work properly carried out together with any other monies due to the contractor under the contract and any loss and expense due to the employer arising from the termination. Taking into account amounts previously paid, the balance must be paid to either employer or contractor as before. If the employer has not acted by the end of the six-month period, the contractor is entitled to require the employer to state in writing whether the work is or is not to be completed.

If the contractor terminates under SBC, DB, IC or ICD clauses 8.9 or 8.10, the rights and duties of the parties may be summarised as follows:

- The contractor must remove from the site all its temporary buildings, plant, equipment, goods and materials as soon as reasonably possible after termination. It must give its sub-contractors similar facilities to remove their property.
- If there is a CDP, the contractor must provide two copies of the design documents prepared.
- Clauses requiring payment or release of retention to the contractor cease to apply.
- The contractor must prepare an account as soon as reasonably practi-cable setting out the amount as follows:

  - the total value of the Works at termination;
  - any sum ascertained under the loss and/or expense clause;
  - reasonable cost of removal of its property from site;
  - the cost of materials properly ordered for the Works for which the contractor has paid or is legally bound to pay;

- any direct loss and/or damage caused to the contractor by the termination after taking into account amounts already paid to the contractor under the contract.

The problem for the employer is the direct loss and/or damage to which the contractor is entitled. Among the damages to be included is the loss of profit which the contractor would have received had the contract proceeded to completion.[21] If the termination occurs after only a few weeks of a long contract, the employer will find this item particularly expensive. Of course, the contractor must demonstrate that if the contract had continued, it is likely that it would have made a profit. Depending on the economic climate, this sometimes might prove a problem to many contractors.

Where termination occurs by notice from either party under SBC, DB, IC or ICD clause 8.11 or under DBC or DB schedule 3 paragraph C4.4 or IC/IOC schedule 1 paragraph C4.4 or by the employer under SBC, DB, IC or ICD clause 6.10.2.2, the rights and duties of the parties are the same as if the contractor had terminated, except that the contractor has no right to any loss and/or damage arising from the termination unless the event concerns damage to the Works by an insured risk due to the employer's negligence.

The consequences of termination under MW and MWD are similar but somewhat shorter. If the employer terminates, everything covered in the other contracts is included except:

- There is no provision for the contractor to be required by the architect to remove temporary buildings, etc. from site.
- There is no provision for the contractor to provide such CDP drawings as have been prepared.
- The contractor cannot be required to assign any sub-contracts to the employer.

If the contractor terminates, everything covered in the other contracts is included except:

- The contractor is not required to remove temporary buildings, etc. from site.
- Where there is a CDP, the contractor is not required to provide such design drawings as have been prepared.
- The reasonable costs of removal of the temporary buildings, etc. are not to be taken into account in preparing the account.

In conclusion, the following basic points should be noted:

- MW and MWD termination provisions are substantially changed and improved from the MW 98 provisions.

- A party contemplating termination should take advice. There is a grave danger that a party trying to terminate will thereby commit a breach entitling the other party to terminate instead.[22]
- From the employer's point of view, termination is always more expensive – in time and money – than persevering to the completion of the contract with the original contractor. This is true no matter who terminates.
- Termination should not be lightly threatened. In more ways than one it is the last resort.

References

1 Westminster Corporation v J. Jarvis & Sons (1970) 7 BLR 64; W. Nevill (Sunblest) Ltd v Wm Press & Son Ltd (1981) 20 BLR 78.
2 Emson Eastern (in receivership) v EME Development (1991) 55 BLR 114.
3 Token Construction Co. Ltd v Charlton Estates Ltd (1973) 1 BLR 50.
4 William Tomkinson & Sons Ltd v Parochial Church Council of St Michael (1990) 6 Const LJ 319.
5 Simplex Concrete Piles Ltd v Borough of St Pancras (1958) 14 BLR 80.
6 William Tomkinson & Sons Ltd v Parochial Church Council of St Michael (1990) 6 Const LJ 319.
7 Pearce and High v John P. Baxter and Mrs A. Baxter [1999] BLR 101.
8 Skanska Construction (Regions) Ltd v Anglo-Amsterdam Corporation Ltd (2002) 84 Con LR 100.
9 English Industrial Estates Corporation v George Wimpey & Co. Ltd (1972) 7 BLR 122.
10 Skanska Construction UK Ltd v Egger (Barony) Ltd [2004] EWHC 1748 TCC.
11 Davis Contractors Ltd v Fareham UDC [1956] 2 All ER 145.
12 Thomas Feather & Co. (Bradford) Ltd v Keighley Corporation (1953) 52 LGR 30.
13 Photo Production v Securicor Transport [1980] 1 All ER 556.
14 Greater London Council v Cleveland Bridge & Engineering Co. Ltd (1986) 8 Con LR 30.
15 West Faulkner Associates v London Borough of Newham (1995) 11 Const LJ 157.
16 John Jarvis Ltd v Rockdale Housing Association Ltd (1985) 5 Con LR 118.
17 D. R. Bradley (Cable Jointing) Ltd v Jefco Mechanical Services Ltd (1988) 6-CLD-07-1; C. J. Elvin Building Services v Noble (2003) CILL 1997.
18 Nash Dredging Ltd v Kestrel Marine Ltd [1986] STL 62.
19 Hounslow London Borough Council v Twickenham Garden Developments Ltd (1970) 3 All ER 326.
20 Kong Wah Housing v Desplan Construction (1991) 2 CLJ 117; Chermar Productions v Prestest (1991) 7 BCL 46.
21 Wraight Ltd v PH&T (Holdings) Ltd (1968) 8 BLR 22.
22 Lubenham Fidelities & Investments Co. Ltd v South Pembrokeshire District Council and Wigley Fox Partnership (1986) 6 Con LR 85.

8   Dispute resolution

8.1 General

Settlement of disputes is dealt with in clause 9 of SBC, DB, IC and ICD and in clause 7 and schedule 1 of MW and MWD. It is simply another of the ways in which the Minor Works Contracts fail to conform with the standards established for the other JCT contracts.

All the contracts now make express provision for the parties to agree to resolve any dispute by mediation (SBC, DB, IC and ICD clause 9.1, MW and MWD clause 7.1). Formerly, it was included as advice in a footnote. There is little point including, in any contract, a provision that the parties can agree something, because it is open to the parties to agree virtually anything. Therefore, this clause is, at best, superfluous or, at worst, misleading. A contract should set out the parties' rights and duties and related procedures. There is no room for this kind of unnecessary clause.

8.2 Adjudication

The Housing Grants, Construction and Regeneration Act 1996, section 108, gives the right to any party to a construction contract to submit disputes to an independent adjudicator. It is clear that the system is very popular and it is expected that most disputes will be resolved in this way. A considerable body of case law has been generated. Section 108 also provides that the contract must:

- enable a party to give notice of adjudication at any time;
- provide a timetable to secure an appointment and referral to the adjudicator within seven days of the initial notice;
- require the adjudicator to make a decision in twenty-eight days from referral;
- allow the adjudicator to extend the period as agreed by both parties or for fourteen days if just the party who started the procedure agrees;
- impose a duty on the adjudicator to act impartially;
- enable the adjudicator to investigate the facts and relevant law without being requested to do so.

The adjudicator's decision will be binding until otherwise agreed by the parties or until the matter is referred to arbitration or litigation (if there is no arbitration agreement). Neither the adjudicator nor the adjudicator's employees or agents will be liable for actions or omissions done as part of the adjudication.

As and when the referral to adjudication is made, both it and any accompanying documents that are sent with it to the adjudicator must simultaneously be copied to the other party. Whereas no particular form or content is dictated for the notice to refer, the same cannot be said of the referral itself, which must clearly set out:

- particulars of the dispute or difference; *and*
- a summary statement of the contentions relied upon; *and*
- a statement of what remedy or relief is being sought.

The contract gives the adjudicator wide powers to set an agenda and the discretion to take the initiative in ascertaining the facts and the law in relation to the matter(s) referred. Moreover, the adjudicator may:

- apply his or her own knowledge and/or expertise;
- open up, reviewing or revising any opinions, notices decisions and the like previously given under the contract;
- visit the site or any other relevant premises used for the preparation of work in connection with the contract and/or call on the parties to carry out testing or opening up of work, or further testing or further opening up;
- take technical or legal advice, and/or, after notice to that effect to the parties, make enquiries of the parties' employees or other representatives.

Once the referral is made then the non-referring party has a right of reply. It should be noted that a contractual right of reply is not a strict requirement of the Act. The Act requires merely that the referral must be made within seven days of notice to refer and says nothing about the non-referring party's right of reply, let alone when that reply should be given or what it should contain. Without doubt a right of reply must exist, and each of the contracts under consideration gives seven days from the date of referral for the response to be made if desired.

Unless and until agreement is reached to be bound, or until the adjudicator's decision is ratified or reversed in arbitration or litigation, the parties are bound by it. They must give effect to it, and if they fail to do so, legal proceedings may be begun in order to secure such compliance, even where all disputes and differences are agreed under the contract to be referred to arbitration as opposed to litigation. It is clear that the courts have a policy of enforcing adjudicators' decisions whether they are good or bad.[1]

In practice, an adjudicator's decision may be resisted only if the adjudicator has acted outside his or her jurisdiction[2] or has failed to observe the rules of natural justice.[3] However, the rules of natural justice are not applied as strictly to adjudication as to other, finally binding, methods of dispute resolution.[4]

The adjudication provisions are to be found in clause 9.2 of SBC, DB, IC and ICD and clause 7.2 of MW and MWD.

There is provision in each of the six contracts to allow an adjudicator to be named in the contract. There is also provision for an adjudicator nominating body. There are disadvantages with both systems. A person with impeccable credentials can be named in the contract, but that person may not be suitable for the particular type of dispute which occurs. On the other hand, a nominating body may appoint as adjudicator someone who turns out to be less than ideal. It is unfortunate that, despite the vetting procedure and criteria adopted by these bodies, some of the adjudicators one encounters are deficient in basic legal knowledge and in the ability to strip away the floss and focus on the real issues in the adjudication. Indeed, there are some adjudicators who believe that they are entitled to reach their decisions by 'gut instinct' rather than by applying proper legal principles. Because adjudication is rough justice and need not be finally binding, the courts will enforce an adjudicator's decision even if it is wrong in fact and law.

The parties need not be legally represented and, if the dispute concerns valuation or quality of work, legal representation may not be essential. However, it should be realised that most disputes have a legal aspect, and the employment of a legal adviser to frame the arguments in the Referral or the Response may be wise.

Instead of setting out its own procedure, as was the case in the JCT 1998 contracts, the adjudication procedure adopted in each contract is the Scheme for Construction Contracts (England and Wales) Regulations 1998 (the 'Scheme').

It is important to understand that adjudication cannot be excluded from a construction contract (except of course in the case of residential occupiers). Adjudication is not an alternative to arbitration or litigation. Arbitration or litigation are the finally binding options. Therefore, each of these JCT contracts must be completed so that the parties may choose either adjudication (for a quick but not necessarily final decision) or one of either arbitration or litigation (legal proceedings).

8.3 Arbitration

Arbitration is the time-honoured system of settling disputes. In essence it happens when two parties who are in dispute about something agree between themselves to ask an independent third party to settle the matter between them and they agree to abide by the decision of the third party.

Arbitration as understood in regard to building disputes has tended to become somewhat cumbersome and peopled by large numbers of solicitors, barristers and expert witnesses. In serious cases involving very large amounts of money it is perhaps understandable why this should be so. It does not seem to be generally appreciated, however, that arbitration can be a relatively quick process if the dispute is minor and dealt with on the basis of documents only. Both parties can agree that legal advisers will not be involved and a settlement can be achieved with the minimum of fuss.

The main advantage of arbitration over litigation in the courts is privacy. No one but those immediately involved need know the details of the dispute and the arbitrator's decision (known as the award). Another advantage is that the parties can choose an arbitrator with technical knowledge who understands, better than a judge, the day-to-day problems of the construction industry. Arbitration can be quicker than litigation provided both parties are anxious to proceed. If there is to be a hearing, it can be arranged in a place and at a time to suit both parties.

The disadvantage of arbitration is that the arbitrator, unlike a judge, has very limited power to move events on if one of the parties is determined to go slow. Arbitration can then become very expensive.

Arbitration is governed by the Arbitration Act 1996. Where two parties have entered into a contract which provides that disputes are to be settled by arbitration, it effectively prevents one party from deciding to pursue the dispute through the courts. The other party will be granted a stay of legal proceedings to allow arbitration to take place. The courts used to have a discretion under certain circumstances to allow litigation to proceed, but under the 1996 Act there is no exercise of discretion, and the only ground to allow litigation to continue and refuse a stay would be if the arbitration agreement were void or inoperative.

It is unusual for an Act of Parliament to set out principles but, importantly, this Act sets out the underlying principles in section 1. These principles are the basis under which the Act will be interpreted. The principles are as follows:

- The object is to obtain the fair resolution of disputes by an impartial tribunal without unnecessary delay or expense.
- The parties should be free to agree how their disputes are resolved, subject only to the public interest.
- The courts should not intervene except as provided by the Act.

One important point stems from this. Many of the provisions are non-mandatory, which means that the parties can exclude them if they so agree. Note that any such agreements must be in writing to be effective. It is vitally important that parties decide carefully which of the non-mandatory provisions, if any, are not going to be adopted. The following are some of the more interesting points:

- The arbitrator must act fairly and impartially.
- The arbitrator must decide how the arbitration is to be run, including the manner of submissions and evidence. If the arbitrator thinks fit, he or she can adopt an inquisitorial role and take the lead in asking questions to find out the facts.
- The arbitrator may order inspection, photography, custody, sampling or experiments on property owned by or in the possession of either party.
- The Act now puts a positive obligation on the parties to pursue the claim without delay and to attend hearings.
- The arbitrator can now order the parties to do things and order them to perform their side of a contract.
- The arbitrator can now award simple or compound interest as seems appropriate.
- The arbitrator can now 'cap' the costs that can be recovered in an arbitration. This should dissuade anyone who is intent on going to arbitration to settle a score rather than genuinely to settle a dispute. This power may encourage parties to exercise restraint and avoid unnecessary expense.

Although agreements to arbitrate must still be in writing, 'agreement in writing' is given a very broad interpretation so that virtually any form of record, whether signed or not, will qualify.

If both parties have signed a standard form contract containing an arbitration clause, there is little doubt that they have entered into a binding agreement to arbitrate; but what is the position if one party has simply written to the other confirming agreement to enter into a contract on standard terms (e.g. SBC)? An arbitration clause is really a separate contract existing alongside the main contract and capable of existing after the main contract has ended. Therefore, if it is intended to incorporate the arbitration clause, it is important to refer to it in clear words.[5] Failure to do so, however, is not necessarily fatal.[6]

There is an arbitration clause in SBC, DB, IC and ICD article 8 and clauses 9.3–9.8. MW and MWD deal with arbitration in article 7, clause 7.3 and schedule 1. The arbitration provisions in all these contracts are virtually identical. It should be noted that if arbitration is the procedure of choice, it must be so stated in the Contract Particulars. Otherwise, litigation is now the default option. This appears to be a bad decision on the part of JCT and overturns a long-established position which used to make arbitration the default procedure.

SBC, DB, IC, ICD, MW *and* MWD

The most important point is that a dispute or difference must have arisen between the parties, i.e. the employer and contractor. The architect is not a party to the contract and, therefore, cannot be a party in the arbitration proceedings, although the architect can be a witness. When exactly a dispute has arisen will be something which should be obvious but, basically, the two parties must be in disagreement.[7]

The subject matter of the dispute can be the interpretation of contract provisions or any other matter arising in connection with the contract, including items left to the architect's discretion; withholding of certificates, adjustment of the contract sum, unreasonable withholding of consent or rights and liabilities of the parties. In fact, virtually any kind of dispute can be referred, provided it bears a relation to the contract.[8]

The party seeking arbitration must write and give the other party notice of reference to arbitration. If they fail to agree on a mutually acceptable person within fourteen days of the initial request, either party may write to whomever is named in the Contract Particulars as the appointor. The appointor will be either the President or the Vice-President of the Royal Institute of British Architects (RIBA), of the Royal Institution of Chartered Surveyors (RICS) or of the Chartered Institute of Arbitrators (CIArb). If it becomes necessary to take this latter course, there is a form to be completed and a fee to be paid. It is worth noting that an appointment made in this way, which is binding on the parties and the person appointed, is almost impossible to remove.

The arbitrator has wide powers under the Arbitration Act, and the contract gives additional powers to direct measurements and valuations, to ascertain and award any sum which should have been included in a certificate to open up, review and revise any certificate, opinion, decision, requirement or notice, and to determine all matters in dispute as if the certificates, etc. had not been given. The arbitrator has power to rectify the contract so that it accurately reflects what the parties agreed. This enables the arbitrator to correct errors and certain specific kinds of mistake. It has been held that an arbitrator always had this power. The new provision simply recognises the point expressly.

The arbitration is to be conducted under the JCT 2005 edition of the Construction Industry Model Arbitration Rules (CIMAR). These are set out in full in a handy booklet (available from the Joint Contracts Tribunal or downloadable from www.jctcontracts.com). An important feature is a choice of procedure. The procedure extends to the taking up of the award. The choice of procedure is welcome: documents only, full procedure or the short procedure with a hearing, which may be quite informal and may even be held on site. It is not open to the parties to agree to dispense with the CIMAR.

The arbitrator is immune from actions of negligence, although the arbitrator may be removed and the award overturned in certain limited circumstances. It used to be the custom for arbitrators to give awards without reasons, but the 1979 Act enabled the parties to request reasons or the courts to order them to be given. It appears to be largely due to the determination of the courts that reasoned awards have not led to a spate of appeals.

The award of the arbitrator is stated to be final and binding. In practice, arbitration awards do tend to be final and binding because, although there is provision in the Arbitration Act 1996 for an appeal to be lodged, it is surrounded by conditions. The appeal must relate to a question of law arising from the award. There is no appeal from facts decided by the arbitrator. All the parties must consent to the appeal or the High Court must give permission.

The courts are reluctant to give permission to appeal to the High Court, and appeals to the Court of Appeal are virtually impossible unless the matter involves something of public importance.[9] The contracts state that the parties agree and consent that either of them may apply to the courts to decide any question of law arising in the course of the arbitration or appeal to the High Court on a question of law arising from the award or apply to the High Court for a decision on a point of law which arises during the arbitration. This appears to overcome the necessity of seeking permission from the court itself.[10]

English law is to be used in deciding any point under the contract (SBC, IC and ICD clause 1.12, MW and MWD clause 1.7 and DB clause 1.11). This is the case irrespective of the nationality or homes of the parties or the situation of the Works. It is important to settle this point, because in the case of an arbitration between two parties of differing nationalities over a contract in a third country, the position could become complicated. It is always open to the parties to agree that a different system of law is to be applied if they so wish, and a suitable amendment must be made in that case.

Reference to arbitration may be made at any time. As a matter of practice, in view of the relatively short contract periods envisaged by most contracts let on MW/MWD terms, arbitrations are likely to take place after practical completion under those contracts. The 1998 edition of the CIMAR is to be used.

8.4 Litigation

Each contract now has provision for litigation (referred to in the contracts as 'legal proceedings') instead of arbitration. It used to be thought that the arbitrator's powers under the contract were wider than that of the court. For example, the arbitrator had the power to open up and review the architect's decisions. That view was largely based on a case[11] which has been overruled in the House of Lords.[12] Litigation is now the default option and it will apply unless arbitration is expressly stated to apply in the Contract Particulars.

8.5 Points to note

The arbitration clause in a contract will survive the contract itself to allow the arbitrator, for example, to decide whether the contract has in fact come to an end and to determine the consequences for the parties.[13]

The majority of references to arbitration are settled or fade away before being determined by the arbitrator. If the reference is not to be decided on the basis of documents only, a hearing will be involved. The parties will have exchanged claims, defences and counter-claims, discovery of documents will have taken place and expert witnesses will have reported. The whole process tends to be very expensive.

The moral seems to be: do not resort to arbitration unless there really is no alternative. If arbitration is the answer, keep it as simple as possible; when the award arrives it will be final and binding. At that point, the losing party invariably wishes that settlement proposals had been accepted and the costs and mental trauma avoided.

Alternative Dispute Resolution (ADR) is commonly referred to as the modern method of resolving disputes. ADR is essentially a dispute resolution technique which depends upon the willingness – indeed, the determination – of the parties to settle the dispute. If both parties wish to use ADR techniques, there is nothing to prevent their doing so and the courts actively require their use.

References

1 Macob Civil Engineering Ltd v Morrison Construction Ltd [1999] BLR 93.
2 Fastrack Contractors Ltd v Morrison Construction Ltd [2000] BLR 168.
3 Discain Project Services Ltd v Opecprime Developments Ltd [2001] BLR 285.
4 Straume (UK) Ltd v Bradlor Developments (1999) CILL 1520.
5 Aughton Ltd v M. F. Kent Ltd (1991) 57 BLR 1.
6 Roche Products Ltd v Freeman Process Systems (1996) 80 BLR 102.
7 Hayter v Nelson and Others (1990) *The Times* 29 March 1990.
8 Ashville Investments Ltd v Elmer Contractors Ltd (1987) 37 BLR 55.
9 BTP Tioxide Ltd v Pioneer Shipping Ltd and Armada Marine SA. The Nema [1981] 2 All ER 1030.
10 Vascroft (Contractors) Ltd v Seeboard plc (1996) 52 Con LR 1.
11 Northern Regional Health Authority v Derek Crouch Construction Co. Ltd (1984) 26 BLR 1.
12 Beaufort Developments (NI) Ltd v Gilbert Ash NI Ltd (1998) 88 BLR 1.
13 Crestar Ltd v Michael John Carr and Joy Carr (1987) 37 BLR 113.

Table of cases

Clause index

Subject index

Construction Contracts Questions and Answers
David Chappell

Construction law can be a minefield of complications and misunderstandings in which professionals need answers which are pithy and straightforward but also legally rigorous. In *Construction Contracts: Questions and Answers*, specialist in construction law David Chappell answers architects' and builders' common construction contract questions.

Questions range in content and include:

- extensions of time
- liquidated damages
- loss and/or expense
- practical completion
- defects
- valuation
- certificates and payment
- architects' instructions
- adjudication and fees.

Chappell's authoritative and practical advice answers questions ranging from simple queries, such as which date should be put on a contract, through to more complex issues, such as whether the contractor is entitled to take possession of a section of the work even though it is the contractor's fault that possession is not practicable.

In answering genuine questions on construction contracts, Chappell has created an invaluable resource on which not only architects, but also project managers, contractors, QSs, employers and others involved in construction can depend.

July 2006: 216×138 mm: 240pages
Paperback 978-0-415-37597-9

Construction Contracts
John Murdoch and Will Hughes

The fourth edition of this unparalleled text has been thoroughly revised to provide the most up-to-date and comprehensive coverage of the legislation, administration and management of construction contracts.

Introducing this key topic in construction law and management, this book provides students with a one-stop reference for construction contracts.

Significant new material covers:

- procurement
- tendering
- developments in dispute settlement
- commentary on all key legislation, case law and contract amendments

In line with new thinking in construction management research, this fourth edition of an authoritative guide is essential reading for every construction undergraduate.

March 2007: 234×156 mm: 416 pages
Hardback 978-0-415-39368-3
Paperback 978-0-415-39369-0

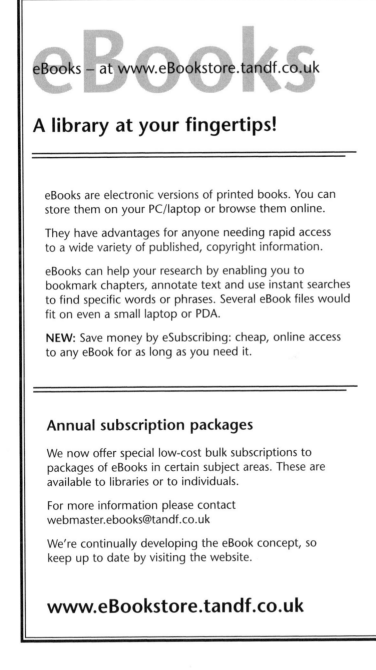